Treatment and Valorisation of Saline Wastewater: Principles and Practice

Treatment and Valorisation of Saline Wastewater: Principles and Practice

Anuska Mosquera Corral, Ángeles Val del Río and José Luis Campos Gómez

IWA PUBLISHING

Published by **IWA Publishing**
Republic – Export Building, 1st Floor
2 Clove Crescent
London E14 2BE, UK
Telephone: +44 (0)20 7654 5500
Fax: +44 (0)20 7654 5555
Email: publications@iwap.co.uk
Web: www.iwapublishing.com

First published 2021

British Library Cataloguing in Publication Data
A CIP catalogue record for this book is available from the British Library

ISBN: 9781789060638 (paperback)
ISBN: 9781789060645 (eBook)

Contents

Authors

Anuska Mosquera Corral
CRETUS, Department of Chemical Engineering,
Universidade de Santiago de Compostela (USC),
15782 Santiago de Compostela, Galicia, Spain.
anuska.mosquera@usc.es

Ángeles Val del Río
CRETUS, Department of Chemical Engineering,
Universidade de Santiago de Compostela (USC),
15782 Santiago de Compostela, Galicia, Spain.
mangeles.val@usc.es

José Luis Campos Gómez
Faculty of Engineering and Sciences,
Universidad Adolfo Ibáñez (UAI),
Avda. Padre Hurtado 750, 2503500,
Viña del Mar, Chile.
jluis.campos@uai.cl.

Acknowledgements

The elaboration of this book in the USC was supported by the Spanish Government (AEI) through the TREASURE project [CTQ2017-83225-C2-1-R] co-funded by FEDER (UE) and, in the UAI, by the Chilean Government through the Projects FONDECYT 1200850 and CRHIAM Centre grant number ANID/FONDAP/15130015. Anuska Mosquera Corral and Ángeles Val del Río belong to the Interdisciplinary Research Center in Environmental Technologies (CRETUS) and to a Galician Competitive Research Group (GRC), the latter programme co-funded by FEDER (UE) as well. José Luis Campos Gómez thanks Carlos Jérez, dean of the Faculty of Engineering and Sciences (UAI), for the reduction of the teaching load granted to promote his collaboration in the writing of this book.

Chapter 1
Introduction

ABSTRACT

Salinisation of freshwater occurs around the world due to anthropogenic activities associated with urban and industrial activities. These activities include groundwater abstraction for potable water supply to levels that favour seawater intrusions in coastal areas, the use of decalcifying products to prevent damage to appliances and the utilisation of salt media in industrial processes. These uses of water produce saline wastewater which is subjected to cleaning treatments that do not include salt removal. Thus, treated wastewater is reintroduced to the environment with salt levels that reduce its quality and make its further utilisation difficult. For this reason, an evaluation of the sources of wastewater with salt concentrations (e.g. NaCl) in the range from 1,300 (moderately saline) to 28,800 mg/L (very highly saline) will be provided in this section. Characteristics and compositions will be described for urban and industrial wastewater. The specific problems associated with the presence of salts will be presented and discussed.

Keywords: Environmental impact, fertigation, freshwater, industrial wastewater, reuse, salinisation, seawater, treatment

1.1 SALINISATION

Dissolved salts are present naturally in water and are necessary as they contribute to maintaining the health and vitality of organisms. Albeit this beneficial effect occurs only if the concentrations of these salts remain below certain levels. Sodium salts are the dominant ones (mainly as NaCl) in soils and groundwater, but salts of other

Author(s): Anuska Mosquera Corral, Ángeles Val del Río and José Luis Campos Gómez
doi: 10.2166/9781789060645_0001

cations such as calcium, magnesium, iron, boron, sulphate, carbonate, and bicarbonate are also found in different concentrations, mainly depending on the specific locations. If salt concentrations exceed the limiting values this can damage the organisms of the ecosystem. Nowadays, due to human activities (urban, industrial or irrigation uses), a large amount of the water resources in the world are experiencing an increase in salinity. This salinisation of water bodies becomes a problem when it reaches a level that impacts on health, agricultural production, environmental ecosystems equilibrium, and economic welfare.

The salinisation of continental water is directly related to the salinisation of soil. Although the idea that soil salinisation occurs mainly in arid and semi-arid regions is generally widespread, its effects are present all over the world. Major water salinity problems have been reported in the USA, Australia, India, Argentina, Sudan, Israel and many countries in Central Asia (Shtull-Trauring *et al.*, 2020; WWAP, 2017). Water and soil salinisation are classified as primary or secondary. Primary salinisation refers to salinisation processes mobilising natural salts (naturally present in the environment), while secondary salinisation refers to salinisation processes increased or induced by human activity (also called anthropogenic salinisation).

Primary salinisation is a complex process involving the movement of salts and water in soils during seasonal cycles and their interactions with groundwater. While rainfall, aeolian deposits, mineral weathering, and stored salts are the sources of salts, surface and groundwater can redistribute these accumulated salts (Rengasamy, 2006). Furthermore, seawater intrusion into land, due to groundwater overdraft and an increase of the seawater level due to climate change or in recent tsunami-affected regions, can deposit huge amounts of salts in the soils of coastal lands (Flood & Cahoon, 2011).

However, although the groundwater or surface water salinisation is a function of natural phenomena, it is exacerbated by anthropogenic factors, in secondary salinisation. Human activities such as agriculture and livestock farming, as well as the production of domestic sewage (in some regions water softeners and salts are added to prevent calcification in water-using appliances), can be highlighted as the most important contributors to salinisation phenomena. In all these activities saline wastewater is produced. Moreover, utilisation of seawater as a supplement to water supply in situations where high quality freshwater is unnecessary also increases the amount of saline wastewater. In agriculture the use of poor-quality irrigation water contributes to salt accumulation in irrigated soils. As an example, Martín-Queller *et al.* (2010) observed that in a semi-arid Mediterranean region (Flumen River, Spain) urban activity and livestock farming increased the concentration of salts in the order of 240–541 μS/cm. Specifically, these authors measured the increase of nitrate concentration from 8.5 to 20.8 mg NO_3^-/L during irrigation months, associated with high levels of irrigation return flows. Additionally, urban surrounding areas contributed to an increase of phosphorus concentrations from 0.19 to 0.42 mg P/L in the river. Finally, their data also indicated that salinisation of soils, subsoils, surface water, and groundwater can be an unwelcome result of the application of pig manure for fertilisation which increased sodium concentrations from 77.9 to 138.6 mg Na^+/L.

In certain cold regions, the direct application of salts to roadways and sidewalks to melt ice and snow contributes significantly to this phenomenon as well. For example, Meriano *et al.* (2009) reported that in Ontario (Canada) 50% of the salt applied to paved road surfaces is conveyed to catchment streams and Frenchman's Bay Lagoon via surface runoff. The remaining 50% enters the subsurface as aquifer recharge and migrates towards Frenchman's Bay Lagoon. As a result, stream water quality is seriously degraded year-round due to the influx of salt from both runoff and baseflow. Therefore, chloride concentrations throughout the watershed consistently exceed the Ontario Provincial Water Quality Objective of 250 mg Cl^-/L.

As stated here, salinisation of water is occurring all over the world and will progressively decrease the water resources available for food production and freshwater uses. For this reason, it is important to define strategies to preserve the existing freshwater reservoirs and restore the conditions of produced wastewater to discharge non- or low-saline treated effluents to the environment.

1.2 SALINITY QUANTIFICATION

As salt concentration in water can be due to the presence of very different compounds (including a wide variety of ions and cations), the common parameter used to quantify the salinity is electrical conductivity (EC) measured at 25°C. The EC does not identify the dissolved salts but gives a reliable indication of salinity levels via an easy and inexpensive measurement. The EC is measured in the International System of Units (SI) as siemens per metre (S/m), although it can be found in the literature expressed in other units (mS/cm, dS/m, mmhos/cm and µS/cm). The salinity can also be measured as total dissolved solids (TDS, in mg/L); however this measurement is more time consuming and for this reason the TDS value can be obtained indirectly as a function of EC through the use of conversion factors.

Therefore, in the literature salt concentration appears expressed sometimes as EC and other times as TDS. The equivalence between these units is presented (Table 1.1), to help with the understanding and comparison of the parameters used to express salinity throughout this book.

Table 1.1 Equivalence between units for salinity.

EC (S/m)	EC (mS/cm, dS/m, mmhos/cm)	EC (µS/cm)	TDS (mg/L)
0.1	1.0	1,000	TDS = K · EC (µS/cm)[a]

[a]Variable values of K (0.50–0.75) depending on the EC value (Walton, 1989).

1.2.1 Classification of saline water

Water resources can be classified depending on their salt concentration according to the rank established by the Food and Agriculture Organization of the United Nations (FAO) (Rhoades *et al.*, 1992) that is summarised in Table 1.2. This classification helps with understanding the possible uses of saline water and wastewater for crop production, as only very tolerant crops can be successfully produced with water with EC that exceeds 10 mS/cm.

Table 1.2 Classification of saline water according to FAO (Rhoades *et al.*, 1992).

Water Class	Electrical Conductivity (mS/cm)	TDS (mg/L)[a]	Type of Water
Non-saline	<0.7	<450	Drinking and irrigation water
Slightly saline	0.7–2.0	450–1,300	Irrigation water
Moderately saline	2–10	1,300–6,400	Primary drainage water and groundwater
Highly saline	10–25	6,400–16,000	Secondary drainage water and groundwater
Very highly saline	25–45	16,000–28,800	Very saline groundwater
Brine	>45	>28,800	Seawater

[a]Assuming an average value of K = 0.64 for the conversion of EC (as μS/cm) to TDS (as mg/L) (Metcalf & Eddy, 2003).

1.2.2 Salinity in recycled water for irrigation

The increasing trend towards using treated wastewater for irrigation or land application is contributing to the salinisation rise in soils. It was widely reported that this effect does not depend only on the EC of the effluent, but also on other components, such as suspended solids and organic matter, that can alter the hydraulic conductivity and infiltration rates. To measure these effects an index called the Sodium Adsorption Ratio (SAR) is used. The SAR expresses the relative activity of sodium ions in the exchange reactions taking place in the soil. This ratio is an indicator of the suitability of a certain water to be used in land irrigation and, also, a standard diagnostic parameter for the sodicity hazard that it exerts on the soil. Equation (1.1), applied to determine the SAR, considers the relative concentration of sodium referred to the calcium and magnesium ones as:

$$\mathrm{SAR} = \frac{[Na^{+}]}{\sqrt{\left(\frac{[Ca^{+2}] + [Mg^{+2}]}{2}\right)}} \tag{1.1}$$

The suitability of reclaimed water for use in land irrigation as a function of the SAR value can be defined as follows:

- For SAR values below 3.0: no restriction exists.
- For SAR values from 3.0–9.0: care needs to be taken when irrigating sensitive crops. Soils should be sampled and tested every 1 or 2 years to determine whether the water is increasing the sodium content.
- For SAR values over 9.0: water is not suitable for irrigation and risk of severe damage of the soil exists.

1.3 ORIGIN OF SECONDARY SALINISATION

The salinisation of water and soil by natural phenomena is out of the scope of this book and for this reason this section is focused only on the discussion of secondary salinisation causes. Anthropogenic activities which can be highlighted as the major sources of secondary salinisation in water can be classified in three main categories: groundwater overdraft, agriculture practices and wastewater (domestic and industrial) discharges.

Groundwater overdraft: in some locations, groundwater overdraft (excessive water abstraction) has caused the natural groundwater gradient to reverse and allowed seawater to intrude coastal aquifers that historically contained only freshwater. Seawater intrusion can be detrimental to drinking water and irrigation wells and render some areas unsuitable for continued agriculture. To prevent additional seawater intrusion, some communities have installed subsurface barriers and injection wells to restore or at least diminish the salinity of the groundwater. Tularam and Krishna (2009) revised the long-term consequences of groundwater pumping and they found that the majority of the affected areas are coastal regions where population density is high. These authors report examples of seawater intrusion due to groundwater overdraft in different world regions such as South-Central Kansas (USA), Mediterranean and South-Atlantic coastlines (Spain), Israel, Mexico, Chile, Peru, Cyprus, Australia, etc. Zektser *et al.* (2005) evaluated different case studies of the south-western United States where groundwater extraction together with overdraft occurs, to highlight the importance of the development of corresponding environmental regulations. These authors stated that the four main impacts associated with these two activities are reduction of streamflow and lake levels, diminishment or elimination of vegetation, land subsidence and seawater intrusion. One example of seawater intrusion into groundwater occurred in the Korban aquifer (north-eastern Tunisia). In this case, treated domestic wastewater is used to artificially recharge the aquifer level and reduce the salinity in the groundwater (Horriche & Benabdallah, 2020).

Agriculture practices: plants can naturally increase soil salinity as they uptake water and exclude salts. However, the main sources of salinisation in agriculture are irrigation and fertilisation techniques. Application of synthetic fertilisers can increase nitrate concentrations in surface and groundwater, while the use of manure from confined animal facilities, rich in nutrients and other salts, can also increase salinity levels in receiving water bodies. Furthermore, inefficient irrigation and drainage systems cause an excessive water leakage and increase the risk of salinisation and inundation in irrigation areas. Performed estimations indicate that of the annual production of wastewater in the world (approximately 30 million tons) 70% is consumed as an agricultural fertiliser and irrigation source (Barbera & Gurnari, 2018). Additionally, the abusive groundwater overdraft for agriculture purposes contributes to seawater intrusion (Ben-Asher *et al.*, 2002). According to Smedema and Shiati (2002) the salinisation problem provoked by agriculture practices is more severe in arid soils, estimating that from the land dedicated to agriculture 60% corresponds to humid climatic conditions and the remaining 40% to semi-arid regions. In these semi-arid regions, the salinisation is a more severe problem because of two features: the naturally highly saline soil and the very high rates of evapo(transpi)ration. These authors indicate that as crops absorb only a fraction of the salt of the consumed water, irrigation causes these diluted salts to become concentrated. Therefore, they estimate that from 3 to 5 tons of salt are concentrated in soils per irrigated ha, per year.

Wastewater discharges: detergents, water softeners, and some industrial processes contain or use salts. Wastewater collected in wastewater treatment plants (WWTPs) and septic systems is often saltier than the freshwater used in the process it originates from and when released to the environment can increase the salinity of the receiving water bodies. Overwatering of lawns and residential uses can also contribute to augmenting water salinity, as can many industrial processes, such as printing, dyeing, refining, chemical production, mining, curriers, pharmaceuticals, power plants and food processing plants. Data from the United States (Dieter *et al.*, 2018) indicate that of the total water withdrawals in 2015 6% was saline. Of these saline discharges 97% corresponded to the thermoelectric industry, 2% to the industrial sector and 1% to mining. The salinisation potential of domestic and industrial effluents is analysed in more detail in the following sections.

1.3.1 Domestic sewage

The wastewater produced in households contains higher salinity than the supplied freshwater because of the use of water softeners, detergents, cleaning products, soaps, shampoos, etc. For this reason, the EC of used water is generally between 70 and 100 mS/cm higher than that of potable water (HWT, 2020). Additionally, when water source separation is practised some of the obtained streams experience a rise in salinity. This trend is like that observed when seawater is utilised to flush toilets. For example, some coastal cities, such as Hong Kong and

Qingdao, use seawater for fire control, road flushing, toilet flushing and other uses that do not required direct contact with human beings, to reduce the pressure on freshwater (Yiyi Zhao *et al.*, 2020). However, these practices produce saline wastewater effluents which cannot always be treated with conventional processes.

Normally, the conventional wastewater treatment processes are not designed to remove salts, but they can help reducing the salinity to a certain level. Levlin (2010) reported that processes such as the removal of suspended solids, through sedimentation or filtration, have no effect on the EC. However, the biological nitrogen removal contribution to the conductivity decrease is significant, since ammonium nitrogen and alkalinity contribute approximately 33% and 14% of the EC, respectively. For example, it is estimated that if the amount of available alkalinity is high enough (meaning one mole of alkalinity per mole of ammonia) biological nitrogen removal via nitrification followed by denitrification, will give a decrease in conductivity of $842\,\mu S \cdot m^2/g\,N$ (expressed as the ratio between the EC, $\mu S/m$, and N concentration, $g\ N/m^3$). Nevertheless, as the contribution of phosphate ions to EC is approximately 1% of the measured conductivity, phosphorous removal will not result in any decrease in conductivity.

Regarding the EC value and/or salt concentration of domestic sewage, different values are reported depending on the geographical zone where it originates, the consumption habits, and the existence of mixing with industrial effluents (Table 1.3).

Table 1.3 Summary of EC values found in sewage in different geographical zones.

Type of Wastewater	EC (mS/cm)	Geographical Zone	Reference
Raw sewage	0.22–0.37	Warri, Nigeria	(Uwidia & Ukulu, 2012)
Treated sewage[a]	0.75–1.02	East London, South Africa	(Odjadjare & Okoh, 2010)
Raw/treated sewage	0.22–1.78/ 0.15–1.78	Jiaxing City, China	(Yu *et al.*, 2019)
Raw sewage	0.88–1.88	Ciudad de Mexico, Mexico	(Ontiveros-Capurata *et al.*, 2013)
Fresh water/treated sewage	0.6/1.00–2.50	Algarrobo, Spain	(Muñoz-Sánchez *et al.*, 2018)
Raw/treated sewage	1.07/1.04	Tunceli, Turkey	(Tanyol & Demir, 2016)
Raw sewage[a]	3.60	Bangladesh, India	(Karmoker *et al.*, 2018)

[a]Domestic mixed with industrial sewage.

1.3.2 Industrial effluents

The number of industrial sectors that produce effluents where salinity can be an issue is huge. Even in the same industrial facility streams are produced with different compositions in terms of organic matter, nutrients, and salt concentrations, requiring separated treatment solutions. Several industrial sectors where the production of saline effluents is relevant are analysed here. In some of these sectors, such as livestock, although the salt concentration in the effluents is not as high as in other sectors (for example the leather industry), the large flows produced (meat production consumes around 25% of the fresh water used to produce aliments) and the significant number of facilities make the treatment of these effluents, normally applied as fertiliser in agriculture, a significant management problem.

In addition, depending on the industrial sector, the salts present in the effluents are different. As an example, wastewaters rich in sulphate are commonly produced in industries where sulphuric acid or sulphate-rich feedstocks are used, as in the fermentation or sea food processing industry. The presence of sulphate causes in anaerobic digesters formation of H_2S, which is the only compound produced in WWTPs from a salt transformation.

1.3.2.1 Intensive livestock

Concentrated, large-scale livestock production often creates concentrated, large-scale environmental problems. Large industrial farms bring in massive quantities of nutrients in the form of concentrated feed, producing far more waste than can be recycled as fertiliser and absorbed on nearby land. When intensive livestock operations are crowded together, pollution can threaten the quality of the soil, water, air, biodiversity, and ultimately public health. Pollution damage is especially harmful when large numbers of animals are concentrated in sensitive areas around cities or close to water resources. Effluents are commonly discharged into the environment or stored in vast 'lagoons', from which waste may spill or leak into nearby streams and groundwater supplies.

The dominant ions that contribute to the EC of manures are NH_4^+, Na^+, Ca^{+2}, Mg^{+2}, K^+, Cl^-, SO_4^{2-} and HCO_3^-, with variable values (Manitoba, 2015). However, it is important to highlight that variations in the EC values are frequently due to variations in the nitrogen content (as NH_4^+), for this reason the EC is not a good measure of the effects that manures will have on soil salinity when they are applied as fertilisers (Manitoba, 2015). In general manure is classified as moderately or highly saline effluent (Table 1.4).

1.3.2.2 Food processing industry

The growth in world population has an associated increase in food demand, which cannot be satisfied with fresh aliments. Therefore, the consumption of processed food rises. Thus, the associated industrial sector must multiply its production

Table 1.4 Summary of EC values found in different types of manure.

Manure Type	EC (mS/cm)	Class	Reference
Cow	4.84 ± 0.28	Moderately saline	(Santillán *et al.*, 2014)
	9.6	Moderately saline	(Irshad *et al.*, 2013)
Cattle	7.04–7.18	Moderately saline	(Yunchen Zhao *et al.*, 2014)
Poultry	7.04 ± 2.57	Moderately saline	(Li & Lujia, 2007)
	8.3	Moderately saline	(Irshad *et al.*, 2013)
Buffalo	8.6	Moderately saline	(Irshad *et al.*, 2013)
Camel	8.7	Moderately saline	(Irshad *et al.*, 2013)
Goat	10.3	Highly saline	(Irshad *et al.*, 2013)
Pig	9–28	Highly saline	(Manitoba, 2015)

capacity and consequently the volume of wastewater generated augments in proportion. The characteristics of the effluents from the food-processing industry differ depending on the main product processed (milk, vegetables, meat, fish, etc.), and even inside the same sector differences can be found between facilities dependent upon the production process arrangement. Nevertheless, there is a common aspect among these sectors, that is the necessity to move towards more environmentally protective processes in which the resources consumed and the waste generated are minimised. For this reason, the effluents produced become more concentrated (as the use of water is reduced) which results in increasing salinity.

In the food-processing industry, salt can be used for two purposes: to add nutritional components and/or to conserve the aliments and avoid their deterioration with time. Among the industrial facilities of the sector the meat canning, vegetable pickling and fish-processing plants stand out as the highest salt consumers (Lefebvre & Moletta, 2006). In Table 1.5 a summary of EC values found in literature for the effluents from different industries of this sector are presented. Regarding the quantity of water consumed, it is estimated that more than two-thirds of all freshwater abstraction worldwide (and up to 90% in some countries) goes towards food production (including agriculture) (Barbera & Gurnari, 2018). Food-processing activities consume different volumes of the freshwater in the following proportions: meat processing 24%, beverages 13%, dairy 12%, fruits and vegetables 10%, bakery and related products 9%, grain and oilseeds 9%, sugar and confectionary 5%, animal food 5%, seafood 2% and other food 11% (Bustillo-Lecompte & Mehrvar, 2015).

1.3.2.3 Tannery industry

Tanneries are typically characterised as pollution intensive industrial complexes which generate widely variable, high-strength wastewater. The major

Table 1.5 Summary of different EC values found in effluents from the food-processing industry.

Industry Type	EC (mS/cm)	Class	Reference
Distillery industry	0.36	Non-saline	(Noukeu *et al.*, 2016)
Sugar refinery	0.3–0.5	Non-saline	(Noukeu *et al.*, 2016)
Confectionery making	0.54	Non-saline	(Noukeu *et al.*, 2016)
Oil mill/soap factory	0.39–1.41	Slightly saline	(Noukeu *et al.*, 2016)
Biscuit factory	1.07	Slightly saline	(Noukeu *et al.*, 2016)
Dairy industry	1.09	Slightly saline	(Noukeu *et al.*, 2016)
Winery industry	1.22	Slightly saline	(Noukeu *et al.*, 2016)
Beer industry	1.49	Slightly saline	(Noukeu *et al.*, 2016)
Soft drinks factory	2.11	Moderately saline	(Das & Kumar, 2009)
Meat processing industry	2.50	Moderately saline	(Luo *et al.*, 2004)
	3.6–9.6	Moderately saline	(Wu & Mittal, 2012)
Vegetable pickling factory	4.0–6.0[a]	Moderately saline	(Ou *et al.*, 2018)
Fish canning industry	4.73–24.8	Moderately-highly saline	(Cristóvaõ *et al.*, 2015)

[a]Value reported as salt concentration.

contaminants present in these effluents are heavy metals, toxic chemicals, chloride, lime with salts and other pollutants, which are generated in the range of 30–35 L/kg skin processed (Durai & Rajasimman, 2011). Certain streams are hypersaline, such as the pickling ones and the chromium tanning effluents or the soak liquor generated by the soaking of hides and skins that can contain as much as 80 g NaCl/L (Lefebvre & Moletta, 2006). Depending on the stage of the manufacturing process, the EC value can vary between 10 and 295 mS/cm (Chowdhury *et al.*, 2015).

1.3.2.4 Petroleum industry

There are three main sources of wastewater contamination in a refinery (Radelyuk *et al.*, 2019): (1) recycling of sulphurous oil and treatment of petroleum products with alkalis, which give highly concentrated sulphurous alkaline wastewater; (2) complex processing of oil and gas to produce synthetic products which generate wastewater with organic acids, alcohols, phenols, etc.; and (3) processes of desalination and dehydration, which cause the presence of demulsifiers and sulfonaphthones in the wastewater. Therefore, the generated wastewater contains suspended solids, dissolved salts, oil and grease, nitrogen compounds, organic sulphides and other substances (heavy metals, phenols, etc.). Thus, a complex treatment train needs to be applied to comply with discharge requirements. Radelyuk *et al.* (2019) examined the wastewater treatment trains that are normally applied in this sector and proposed the use of the following steps: in the pre-treatment screens, sand traps and equalisation tanks; in the primary treatment oil traps, coagulation-flocculation, dissolved air flotation and electrocoagulation-flocculation; in the secondary treatment activated sludge and moving bed biofilm reactor; finally in the post-treatment wetlands, advanced oxidation processes and membrane technologies (if water is going to be reused).

Regarding the characteristics of the wastewater, not only the presence of high salinities, but also other compounds, are a drawback for biological processes. According to Al-Khalid and El-Naas (2018) the effluent from a petroleum refinery contains average values of: 5.2–6.8 mS/cm (EC), 3,800–6,200 mg/L as TDS, 14.5–16.0 mg SO_4^{-2}/L and 160–185 mg/L as total phenol. Regarding the amount of wastewater generated, these authors indicated that, on average, processing a barrel of crude oil consumes 246–341 litres of water with the subsequent wastewater production.

1.3.2.5 Mining industry

Water is used in the mining sector in different operations and processes such as exploration, mineral processing, (treatment of ore, recovery of metals and the control of dust particles) and closure. Therefore, the effluents generated contain different types of dissolved ions, including Na^+, Ca^{+2}, Mg^{+2}, K^+, Cl^- and SO_4^{2-} among others, that vary depending on the geographical zone (type of soil and climate), the processes applied, etc. The water can be obtained from a variety of sources including allocation from the bulk water infrastructure (third-party water), groundwater (mainly in arid regions), surface water (rainfall and runoff), and even water with low quality or alternative sources of water (Thiruvenkatachari & Su, 2017).

The water consumed and the conductivity of the effluents generated in this sector depend highly on the type of mineral extracted. For example in the extraction of coal a water consumption of 180–220 litres per tonne of coal is estimated with an EC value between 0.5–21.0 mS/cm, while for gold this consumption can account for 250 million of litres per tonne and the EC value of the stream generated is between 96–289 mS/cm (Thiruvenkatachari & Su, 2017).

Therefore, the high volumes of water consumed, and the high salinity of the effluents generated, make the application of treatment processes necessary with the aim to reuse the effluents in the same industry. Foureaux *et al.* (2020) compared the use of two different membrane-based technologies (direct contact membrane distillation process and high temperature nanofiltration process) for the treatment of an effluent from a gold mine (with EC values of 34–56 mS/cm). Although with both systems these authors obtained large pollutants rejection rates (>92.4%), the direct contact membrane distillation was proved to have the lowest operational costs, which is important for full-scale application.

1.4 WATER SALINISATION EFFECTS

Currently, about 7% of the world's land surface and over 20% of the world's agricultural land are threatened by salinisation. The problems associated with water salinisation are diverse and with different consequences. Anthropogenic salinisation disrupts ecosystem functions and services, degrades biological systems, and can negatively affect human health and wellbeing (Schuler *et al.*, 2019). In this chapter these problems are classified into two main categories: effects on living organisms and on economy.

1.4.1 Damaging living organisms

An excess of salinity in water has consequences for living organisms and their health. In the case of human beings, apart from the bad taste, salinised drinking water affects mainly kidney function and can also cause hypertension. Furthermore, urban freshwater salinisation can increase the prevalence of salt tolerant disease vectors such as mosquitoes, and salts can mobilise toxic heavy metals into drinking water, increasing the risks to human health. Salt also negatively affects the health and productivity of livestock, especially pregnant females. Typically, small freshwater invertebrates are the most sensitive to increased salinity, while large vertebrates are generally more tolerant.

In the case of crops and plants it affects their growth and development, especially in the early-growth stages. Salt influences how easily plants can extract water from the soil. Plants either store salt within their cells or try to exclude it. Either way, plant growth and function are reduced, which can ultimately lead to the death of the plant.

Also, ecological effects occur, such as imbalance of ecosystems composition due to the transition to salt-tolerant species, which can even happen with small increases in salinity.

All these effects cause problems in the food production chain (crop cultivation and livestock), thus affecting directly human food availability. Khanom (2016) reported these effects for the region of Bangladesh with data from 2000 to 2009. These authors concluded that yield loss in every year has increased and that to cope with soil salinity the native rice varieties need to be changed to others with high yield in these conditions. Additionally, it was necessary to increase the use of fertiliser and pesticides and crops like oilseed, sugarcane and jute were cultivated discontinuously due to their inability to cope with salinity level. Sharma *et al.* (2016) revised the impact of saline water consumption on different animals and highlight that the presence of ions such as Ca^{2+}, Mg^{2+}, Na^+, Cl^-, SO_4^{2-}, and HCO_3^- may cause harmful effects resulting in poor performance, illness, or even death. Furthermore, they state that tolerance for salinity depends on age, species, requirement, season, and physiological condition. For example, they explain that for sheep the desirable maximum drinking water TDS concentration is approximately 5,000 mg/L, while for cattle it is 2,500 mg/L.

1.4.2 Limiting economic and social development

Salinisation limits economic and social development affecting different sectors such as agriculture and livestock, households and public services, and industrial activities. The consequences in each sector are interconnected resulting in a net reduction of economic development and welfare of the region affected by salinity. Social consequences include the displacement of hundreds of thousands to millions of people (migrations) because there is no adequate water and food for the entire population, but also due to the scarcity of work associated with the reduction in the economic activity (agriculture, livestock and industry). Additionally, these migrations can be indirectly motivated by the reduction in other ecosystem services in which freshwater is necessary (Schuler *et al.*, 2019).

Salinisation worsens the environment and the quality of the soil and the water, therefore it increases costs of food production (crops and livestock), because there are extra costs associated with land and water conditioning, for example ensuring sufficient quality for irrigation and for animal consumption. More specifically, high concentrations of salts in water can damage crops, affect plant growth, degrade drinking water for animals, and damage the equipment necessary for the agricultural and livestock activities.

For households, the poor water quality due to the presence of salts results in the appearance of diseases, which have negative economic and social impacts. For example, the corrosion of copper pipes leads to maintenance cost increases and to high levels of dissolved copper in drinking water, which is harmful to humans. Furthermore, the salinisation of water resources reduces the quality of the drinking water and obligates the installation of costly systems at the treatment plant to remove the excess salt, resulting in significantly increased rates for residential and commercial users. Additionally, the public administration is forced to promote improvements in mandatory regulations to guarantee the welfare of the population.

For industries that use water the production process is more expensive as they need to pre-treat it, for both freshwater and in the case that it is reused in the plant after its initial utilisation. Additionally, it is important to consider that these types of treatments generate new residues (separated salts, brines, etc.) that need to be managed, with the corresponding extra cost for the industry. Furthermore, industrial processes suffer damage to their equipment even though they can use saline water to a certain extent. With high salinity, equipment such as pumps, metallic pipes and tanks are corroded. Salt will also affect the function of devices such as air-conditioners, batteries, boilers, engine radiators, swimming pools, etc.

From the economic point of view, quantifying the costs of salinisation is difficult because it impacts in several different aspects, as mentioned previously. To serve as an example, the distribution of the salinisation costs among the different sectors is presented in the case of the Murray-Darling Basin (Australia) (Wilson, 2004): commerce and industry 21%, households 25%, local government assumes 10%, state government agencies and utilities 11%, dryland agricultural producers 33%, environment and cultural heritage (not valued). In this report the main economic costs of salinisation are grouped in the following categories:

(1) Repair and maintenance costs: for example, because the infrastructures are affected more by corrosion under saline conditions.
(2) Costs from the reduced lifespan of infrastructure: attributed to the earlier replacement than normal because of damage caused by the salinity.
(3) Costs of taking preventive actions: includes the amelioration costs incurred by the community to minimise current and future problems, for example with the use of less corrosive materials which are more expensive or with the installation of extra infrastructures to avoid the scarcity of water by salinisation (desertification). Additionally, they can include the costs associated with research programmes related with the salinity issue.
(4) Operating costs: increase of costs related with the use of additional goods and services which minimise the salinisation, for example, treatment of groundwater for drinking.
(5) The value of income foregone: for example, the reduction of incomes in the agricultural sector, the reduction of the property values because they are located in a salinised area, etc.
(6) Environmental and cultural heritage costs: it considers the negative impacts on the natural environment and cultural heritage.

1.4.3 Impacting the environment

To evaluate the environmental impact of salinity different tools can be applied. Among them the life cycle assessment (LCA) is a methodology widely extended to calculate the environmental impact of a product over its entire life cycle. The LCA results are related to environmental impact categories, which include: abiotic depletion, aquatic ecotoxicity, freshwater acidification, cumulative energy demand, eutrophication, global warming, human toxicity, ozone depletion, photochemical ozone creation, photo-oxidant formation, terrestrial ecotoxicity and water demand. However, salinisation is missing in the range of impact categories of most LCA case studies, resulting in the lack of a comprehensive approach to assess salinisation impacts due to human interventions in the LCA framework. For this reason, there is an increasing interest in developing a new impact category which relates the effect of salinity. Payen *et al.* (2016) reviewed the salinisation impacts in LCA for their consistent integration and concluded that there are four relevant aspects to consider associated with human interventions causing soil and water salinisation (secondary salinisation): 1) land use change, 2) irrigation, 3) brine disposal, and 4) over-reuse of a waterbody. These authors established the relationships between these factors and the impact on human health, ecosystem biodiversity and resource (soil and water) availability for future generations, in order to help to understand the basics to apply salinisation as a category in the LCA.

1.4.3.1 Water quality for reuse

The most valuable fraction of the wastewater is the water itself (Verstraete *et al.*, 2009). Therefore, recovery of water from waste streams is increasing in interest. Additionally, the new concept of circular economy is forcing the implementation of strategies that improve the water use, reducing the utilisation of natural resources and recycling the water as much as possible in the production processes. The application of recycled water can be divided into seven categories including: urban reuse, agricultural reuse, impoundments, environmental reuse, industrial reuse, groundwater recharge/non-potable reuse, and potable reuse, with agricultural water reuse being the dominant application in the world (Shoushtarian & Negahban-Azar, 2020).

Therefore, water reuse is nowadays an option that many countries are applying to solve the scarcity of water and to replenish freshwater resources. The technological development of new processes which are more and more efficient makes water reuse a reality. The benefits of water reuse include among others: the decrease of pressure on the freshwater resources, the reduction of pollutants discharged to water bodies and the availability of a water source which is more reliable than other water sources that directly depend on rainfall. Nevertheless, the different regulations all over the world, at international and national level, which describe the quality requirements for water reuse are mainly focused on human-health. They appear insufficient to limit some of the potentially dangerous pollutants, such as emerging compounds, and present large discrepancies when compared with each other. In addition, Shoushtarian and Negahban-Azar (2020), in a review of 70 different regulations and guidelines from different countries and regions, found that some of the important water quality parameters such as pathogens, heavy metals, and salinity are only considered in some of these regulations. In this revision the authors found only 14 regulations/guidelines containing specific restrictions for salinity, among them the maximum EC threshold value is fixed at approximately 3 mS/cm, in line with FAO guidelines (Table 1.2). A summary of the characteristics of salinity that reclaimed water needs to comply with in these regulations is presented in Table 1.6 (adapted from the review of Shoushtarian and Negahban-Azar (2020)).

Table 1.6 Range of thresholds for parameters related to the salinity found in different regulations/guidelines for water reuse (adapted from Shoushtarian and Negahban-Azar (2020)).

EC (mS/cm)	TDS (mg/L)	SAR	Cl^- (mg/L)	Na^+ (mg/L)
0.2–3.0	450–2,000	3–10	100–2,000	70–300

1.4.3.2 Greenhouse gas (GHG) emissions

More specifically the role of salinity in air pollution due to greenhouse gas (GHG) emissions is still unclear and more research is needed to clarify this aspect. Poffenbarger *et al.* (2011) studied the relationship between methane emissions and salinity in tidal marshes and they concluded that methane emissions are negligible in brackish systems but need to be considered relevant in lower-salinity marshes. Similar conclusions were obtained by Wang *et al.* (2017). These authors studied the impact of seawater concentration on the potential production of CH_4, CO_2 and N_2O. They found that potential CH_4 production was unaffected by salinities from 0 to 7.5‰, but declined significantly at 10‰ and above, while potential CO_2 production was stimulated at intermediate salinities (from 5.0 to 7.5‰), but inhibited by salinities over 15‰. Finally, for the potential of N_2O production they observed that it was unaffected by salinity.

Ardón *et al.* (2018) measured the emissions of CH_4, CO_2 and N_2O in soils exposed to marine salt to assess the salinity effect in these GHG emissions and concluded that regarding CH_4 and CO_2 no relevant increase in the emissions were detected, while in the case of N_2O the simultaneous presence of salt (NaCl) and sulphate decreased in 72% of the N_2O emissions.

Shao *et al.* (2020) studied the effect of salinity (0–2%) on GHG emissions (CH_4, CO_2 and N_2O) in a wastewater treatment system with constructed wetlands. Their main findings were that, as salinity decreases chemical oxygen demand (COD) removal, and as CH_4 and CO_2 emissions are positively correlated with COD removal, the increase of salinity reduces their emissions. Regarding N_2O emissions they observed that salinity is negatively correlated with nitrogen removal, therefore if the salinity reduces the nitrogen removal rate the N_2O emissions increase. Along this line, Tsuneda *et al.* (2005) studied specifically the N_2O emission in the biological nitrogen removal processes for wastewater treatment and concluded that in the nitrification process N_2O emissions increase 2.2 times when the salt concentration increases from 1 to 2%, while in the denitrification process the change in the N_2O emission was negligible (even when salinity increased from 3 to 5%). Globally in the activated sludge process (with nitrification-denitrification) these authors observed that a salt concentration increase from 1.6 to 3.0% increased the N_2O emission from 0.7 to 13.0%. They concluded that salinity inhibits the N_2O reductase activity and, consequently, more emissions are detected in the biological nitrogen removal processes.

1.4.3.3 Other impacts

In previous sections the most relevant impacts that water salinisation provokes were discussed. Nevertheless, as a relevant issue there are many other direct or indirect implications to consider, like its contribution to acid rain and desertification.

Saline wastewater containing sulphur compounds (for example due to seawater use or its intrusion) can produce H_2S in the treatment process (in anaerobic conditions), which is a gas that can be oxidised in the atmosphere into SO_2, with the subsequent transformation into H_2SO_4, contributing to acid rain (Rubright *et al.*, 2017).

The nexus between salinisation and desertification is well known. The largest soil areas affected by salinity and sodicity in the world are in Australia and Asia, most of which evolve to desertification of the regions (Singh, 2009). This effect mainly occurs due to the irrigation practices in semi-arid regions since, once the water evaporates, the salts contained in the fertiliser remain concentrated in the soil. Therefore, the salinity of the soils is one of the problems that contributes the most to their degradation in regions susceptible to desertification.

1.5 CONCLUSIONS

Salinisation is a worldwide problem which, although more severe in some places such as semi-arid regions, needs to be considered as a global concern that impacts lives, economic activities, the environment, and social welfare. Salinisation can be due to natural causes (primary salinisation) or anthropogenic activities (secondary salinisation). The latter occurs due to practices concerning agriculture and livestock, industrial processes, and household habits. All these activities are high freshwater demanders, with consequent wastewater generation. Among the sectors producing saline effluents it is important to highlight that in the case of the food-processing industry, livestock activities and urban sewage, if an adequate treatment is applied the effluents can be reused, for example for irrigation. In fact, agriculture is part of the problem and part of the solution for saline effluents. On the one hand the inadequate irrigation practices are responsible for the increasing salinity in soil and groundwater, but at the same time the reuse of saline effluents (from livestock, the food-processing industry and households) reduces the consumption of freshwater resources and decreases the use of fertilisers. Considering that agriculture is the consumer of 70% of the total water demand, finding alternative water resources to freshwater is important, for example to avoid groundwater overdraft and reduce the risk of seawater intrusion.

REFERENCES

Al-Khalid T. and El-Naas M. H. (2018). Organic contaminants in refinery wastewater: characterization and novel approaches for biotreatment. *Recent Insights in Petroleum Science and Engineering*, Chapter 18, 371–391.

Ardón M., Helton A. M. and Bernhardt E. S. (2018). Salinity effects on greenhouse gas emissions from wetland soils are contingent upon hydrologic setting: a microcosm experiment. *Biogeochemistry*, **140**, 217–232.

Barbera M. and Gurnari G. (eds) (2018). Water Reuse in the Food Industry: Quality of Original Wastewater Before Treatments. In: Wastewater Treatment and Reuse in the Food Industry, SpringerBriefs in Molecular Science. Springer, Cham.

Ben-Asher J., Beltrão J., Costa M., Anaç S., Cuartero J. and Soria T. (2002). Modelling the effect of sea water intrusion on ground water salinity in agricultural areas in Israel, Portugal, Spain and Turkey. *Acta Horticulturae*, **573**, 119–128.

Bustillo-Lecompte C. F. and Mehrvar M. (2015). Slaughterhouse wastewater characteristics, treatment, and management in the meat processing industry: A review on trends and advances. *Journal of Environmental Management*, **161**, 287–302.

Chowdhury M., Mostafa M. G., Biswas T. K., Mandal A. and Saha A. K. (2015). Characterization of the Effluents from Leather Processing Industries. *Environmental Processes*, **2**, 173–187.

Cristóvaõ R. O., Botelho C. M., Martins R. J. E., Loureiro J. M. and Boaventura R. A. R. (2015). Fish canning industry wastewater treatment for water reuse – A case Study. *Journal of Cleaner Production*, **87**, 603–612.

Das M. and Kumar A. (2009). Effluent characterization and different modes of reuse in agriculture-a model case study. *Environmental Science and Pollution Research*, **16** (4), 466–73.

Dieter C. A. A., Maupin M. A. A., Caldwell R. R. R., Harris M. A. A., Ivahnenko T. I. I., Lovelace J. K. K., Barber N. L. L. and Linsey K. S. S. (2018). Estimated Water Use in the United States in 2015. U.S. Geological Survey, circular **1441**.

Durai G. and Rajasimman M. (2011). Biological treatment of tannery wastewater – A review. *Journal of Environmental Science and Technology*, **4**(1), 1–17.

Flood J. F. and Cahoon L. B. (2011). Risks to coastal wastewater collection systems from sea-level rise and climate change. *Journal of Coastal Research*, **27**, 652–660.

Foureaux A. F. S., Lebron Y. A.R., Moreira V. R., Grossi L. B., Santos L. V. S. and Amaral M. C. S. (2020) Technical and economic potential of high-temperature NF and DCMD for gold mining effluent reclamation. *Chemical Engineering Research and Design*, **162**, 149–161.

Horriche F. J. and Benabdallah S. (2020). Assessing aquifer water level and salinity for a managed artificial recharge site using reclaimed water. *Water*, **12**(2), 341, 1–11.

HWT. (2020). Electrical Conductivity of water. https://hwt.co.za/electrical-conductivity-of-water/ (accessed 8 November 2020).

Irshad M., Eneji A. E., Hussain Z. and Ashraf M. (2013). Chemical characterization of fresh and composted livestock manures. *Journal of Soil Science and Plant Nutrition*, **13**(1), 115–121.

Karmoker J., Kumar S., Kumer Pal B., Sajal Sorowar M., Aminur Rahman M., Hafijur Rahaman Khan M. and Islam R. (2018). Characterization of wastewater from Jhenaidah municipality area, Bangladesh: A combined physico-chemical and statistical approach. *AIMS Environmental Science*, **5**(6), 389–401.

Khanom T. (2016). Effect of salinity on food security in the context of interior coast of Bangladesh. *Ocean and Coastal Management*, **130**, 205–212.

Lefebvre O. and Moletta R. (2006). Treatment of organic pollution in industrial saline wastewater: A literature review. *Water Research*, **40**(20), 3671–3682.

Levlin E. (2010). Conductivity Measurements for Controlling Municipal Waste-Water Treatment. Joint Polish – Swedish Reports, **15**, 51–62.

Li X. and Lujia H. (2007). Evaluation of specific gravity and electrical conductivity for determining nutrient concentrations in suspensions of poultry manure. *Polish Journal of Environmental Studies*, **16**(6), 919–925.

Luo J., Lindsey S. and Xue J. (2004). Irrigation of meat processing wastewater onto land. *Agriculture, Ecosystems and Environment*, **103**(1), 123–148.

Manitoba (2015). Properties of Manure. https://www.gov.mb.ca/agriculture/environment/nutrient-management/pubs/properties-of-manure.pdf (accessed 8 November 2020).

Martín-Queller E., Moreno-Mateos D., Pedrocchi C., Cervantes J. and Martínez G. (2010). Impacts of intensive agricultural irrigation and livestock farming on a semi-arid mediterranean catchment. *Environmental Monitoring and Assessment*, **167**(1–4), 423–435.

Meriano M., Eyles N. and Howard K. W. F. (2009). Hydrogeological impacts of road salt from Canada's busiest highway on a Lake Ontario watershed (Frenchman's Bay) and lagoon, City of Pickering. *Journal of Contaminant Hydrology*, **107**(1–2), 66–81.

Metcalf and Eddy (2003). Wastewater Engineering: Treatment and Reuse, 4th edn. McGraw-Hill, New York.

Muñoz-Sánchez D., Bogodist V., García-Cañizares V. M., Frías-Gil D., López-Díaz M. R., Jaime-Fernández E. and Romero-Aranda M. R. (2018). Assessing quality of reclaimed urban wastewater from Algarrobo municipality to be used for irrigation. *Journal of Water Resource and Protection*, **10**(11), 1090–1105.

Noukeu N. A., Gouado I., Priso R. J., Ndongo D., Taffouo V. D., Dibong S. D. and Ekodeck G. E. (2016). Characterization of effluent from food processing industries and stillage treatment trial with *Eichhornia crassipes* (Mart.) and *Panicum maximum* (Jacq.). *Water Resources and Industry*, **16**, 1–18.

Odjadjare E. E. O. and Okoh A. I. (2010). Physicochemical quality of an urban municipal wastewater effluent and its impact on the receiving environment. *Environmental Monitoring and Assessment*, **170**, 383–394.

Ontiveros-Capurata R. E., Diakite-Diakite L., Álvarez-Sánchez M. E. and Coras-Merino P. M. (2013). Evaluación de aguas residuales de la ciudad de México utilizadas para riego. *Tecnologia y Ciencias Del Agua*, **4**(4), 127–140.

Ou Y., Gu J., Ran M., Yang H. and Zhang D. (2018). Asynchronous cultivation and domestication of aerobic activated sludge to treat the pickle wastewater. *IOP Conference Series: Earth and Environmental Science*, **186**(3), 1–7.

Payen S., Basset-Mens C., Núñez M., Follain S., Grünberger O., Marlet S., Perret S. and Roux P. (2016). Salinisation impacts in life cycle assessment: a review of challenges and options towards their consistent integration. *International Journal of Life Cycle Assessment*, **21**, 577–594.

Poffenbarger H. J., Needelman B. A. and Megonigal J. P. (2011). Salinity influence on methane emissions from tidal marshes. *Wetlands*, **31**, 381–842.

Radelyuk I., Tussupova K., Zhapargazinova K., Yelubay M. and Persson M. (2019). Pitfalls of wastewater treatment in oil refinery enterprises in Kazakhstan-a system approach. *Sustainability*, **11**, 1618, 1–20.

Rengasamy P. (2006). World salinization with emphasis on Australia. *Journal of Experimental Botany*, **57**(5), 1017–1023.

Rhoades J., Kandiah A. and Mashali A. (eds) (1992). Chapter 2: Saline waters as resources. In: The Use of Saline Waters for Crop Production, FAO Irrigation and Drainage Paper, **48**, 5–9.

Rubright S. L. M., Pearce L. L. and Peterson J. (2017) Environmental toxicology of hydrogen sulfide. *Nitric oxide*, **71**, 1–13.

Santillán Y., Moreno F., Garcia F. and Sandoval O. (2014). Effect of the application of manure of cattle on the properties chemistry of soil in Tizayuca, Hidalgo, Mexico. *International Journal of Applied Science and Technology*, **4**(3), 67–72.

Schuler M. S., Cañedo-Argüelles M., Hintz W. D., Dyack B., Birk S. and Relyea R. A. (2019). Regulations are needed to protect freshwater ecosystems from salinization. *Philosophical Transactions of the Royal Society B: Biological Sciences*, **374**, 1–9.

Shao X., Zhao L., Sheng X. and Wu M. (2020). Effects of influent salinity on water purification and greenhouse gas emissions in lab-scale constructed wetlands. *Environmental Science and Pollution Research*, **27**(17), 21487–21496.

Sharma A., Tariq P. H., Kewalramani N. and Kundu S. S. (2016). Livestock rearing on saline water. In: Innovative Saline Agriculture, J. Dagar, P. Sharma, D. Sharma and A. Singh (eds), Springer, New Delhi, pp. 475–487.

Shtull-Trauring E., Cohen A., Ben-Hur M., Tanny J. and Bernstein N. (2020). Reducing salinity of treated waste water with large scale desalination. *Water Research*, **186**, 116322.

Shoushtarian F. and Negahban-Azar M. (2020). World wide regulations and guidelines for agriculturalwater reuse: A critical review. *Water*, **12**, 971, 1–58.

Singh G. (2009). Salinity-related desertification and management strategies: Indian experience. *Land Degradation and Development*, **20**, 267–385.

Smedema L. K. and Shiati K. (2002). Irrigation and salinity: A perspective review of the salinity hazards of irrigation development in the arid zone. *Irrigation and Drainage Systems*, **16**, 161–174.

Tanyol M. and Demir V. (2016). Correlations between some operation parameters and efficiency evaluation of domestic wastewater treatment plant in Tunceli (Turkey). *Desalination and Water Treatment*, **57**(58), 28115–28121.

Thiruvenkatachari R. and Su S. (2017) Chapter 5: Aerobic Treatment of Effluents From the Mining Industry. In: Current Developments in Biotechnology and Bioengineering, D.-J. Lee, V. Jegatheesan, H. H. Ngo, P. C. Hallenbeck and A. Pandey (eds), Elsevier, pp. 131–143.

Tsuneda S., Mikami M., Kimochi Y. and Hirata A. (2005). Effect of salinity on nitrous oxide emission in the biological nitrogen removal process for industrial wastewater. *Journal of Hazardous Materials*, **119**(1–3), 93–98.

Tularam G. A. and Krishna M. (2009). Long-term consequences of groundwater pumping in Australia: a review of impacts around the globe. *Journal of Applied Science in Environmental Sanitation*, **4**(2), 161–166.

Uwidia I. E. and Ukulu H. S. (2012). Studies on electrical conductivity and total dissolved solids concentration in raw domestic wastewater obtained from an estate in Warri, Nigeria. *Greener Journal of Physical Sciences*, **3**(3), 110–114.

Verstraete W., Van de Caveye P. and Diamantis V. (2009). Maximum use of resources present in domestic "used water". *Bioresource Technology*, **100**, 5537–5545.

Walton N. R. G. (1989). Electrical conductivity and total dissolved solids – what is their precise relationship? *Desalination*, **72**(3), 275–292.

Wang C., Tong C., Chambers L. G. and Liu X. (2017). Identifying the salinity thresholds that impact greenhouse gas production in subtropical tidal freshwater marsh soils. *Wetlands*, **37**, 559–571.

Wilson S. M. (2004). Dryland and Urban Salinity Costs Across the Murray-Darling Basin. An Overview & Guidelines for Identifying and Valuing the Impacts, Murray-Darling Basin Commission, Canberra.

Wu P. F. and Mittal G. S. (2012). Characterization of provincially inspected slaughterhouse wastewater in Ontario, Canada. *Canadian Biosystems Engineering/Le Genie Des Biosystems Au Canada*, **53**, 6.9–6.18.

WWAP (United Nations World Water Assessment Programme) (2017). Wastewater: The Untapped Resource. In The United Nations World Water Development Report 2017. Paris, UNESCO.

Yiyi Zhao, Zhuang X., Ahmad S., Sung S. and Ni S. Q. (2020). Biotreatment of high-salinity wastewater: current methods and future directions. *World Journal of Microbiology and Biotechnology*, **36**, 1–37.

Yu Q., Liu R., Chen J. and Chen L. (2019). Electrical conductivity in rural domestic sewage: An indication for comprehensive concentrations of influent pollutants and the effectiveness of treatment facilities. *International Biodeterioration and Biodegradation*, **143**, 104719.

Yunchen Zhao, Yan Z., Qin J. and Xiao Z. (2014). Effects of long-term cattle manure application on soil properties and soil heavy metals in corn seed production in Northwest China. *Environmental Science and Pollution Research*, **21**, 7586–7595.

Zektser S., Liáiciga H. A. and Wolf J. T. (2005). Environmental impacts of groundwater overdraft: selected case studies in the southwestern United States. *Environmental Geology*, **47**, 396–404.

Chapter 2

Salinity effects on physical-chemical treatments

ABSTRACT

The aims of the physical-chemical processes applied during wastewater treatment are generally to guarantee the good performance of biological systems and improve the effluent quality. Salinity of wastewater affects its ionic strength and density and, therefore, the efficiency of coagulation-flocculation, sedimentation, dehydration, precipitation and disinfection processes. Specifically, the ionic force influences the interaction between particles present in wastewater and, so, their tendency to aggregate. It also affects the growth rate of microorganisms forming flocs, causing changes in the physical structure of the flocs. The increase of the wastewater density negatively affects the settling process. Saline conditions decrease the effective concentration of sparingly soluble salt ions, due to the promotion of electrostatic interactions or the formation of complexes, which increase the solubility of these salts. Inorganic ions can also act as promoters and/or scavengers of radicals during disinfection with UV and advanced oxidation processes.

Keywords: Bubbles coalescence, DLVO theory, flocs structure, ion exchange, particles aggregation

2.1 INTRODUCTION

Both urban and domestic wastewater contain a variety of pollutants that are usually removed by applying physical, chemical and/or biological processes (Figure 2.1).

Author(s): Anuska Mosquera Corral, Ángeles Val del Río and José Luis Campos Gómez

doi: 10.2166/9781789060645_0023

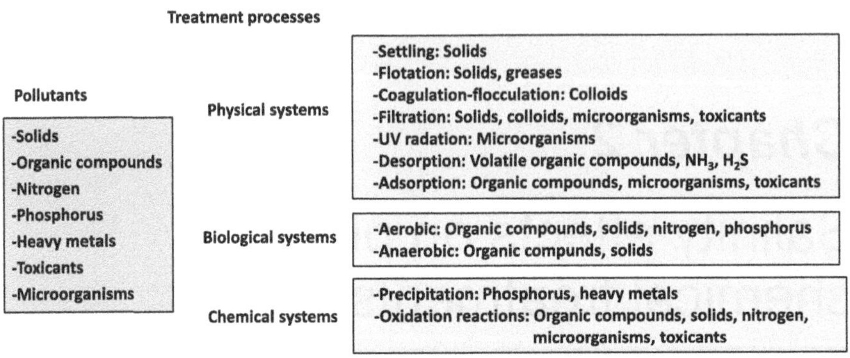

Figure 2.1 Different processes generally applied to remove pollutants from wastewater.

Since none of these processes is capable of individually obtaining the effluent quality required by the discharge regulations, several of them should be usually applied in series to achieve the desired effluent quality. The treatment train that is applied will depend both on the characteristics of the wastewater and the discharge limits. In the case of urban waters, its composition is relatively homogeneous throughout the world, so treatment systems are practically standardized (Figure 2.2a). In the case of industrial wastewater, the characteristics of waste streams vary greatly from one productive sector to another, so a specific treatment plant design is needed for each case (Figure 2.2b). Urban wastewater and most industrial wastewaters contain mostly biodegradable pollutants, so their removal is carried out by means of biological processes. However, before and after these biological processes, chemical-physical processes are applied in order to maintain their good performance and improve the effluent quality, respectively. In the case that industrial wastewater contains non-biodegradable and/or toxic compounds, only chemical-physical treatment systems are applied.

The effect of salinity on the different biological processes that are generally applied to remove organic, nitrogen and phosphorus compounds will be discussed in the next chapter of the book while this chapter will describe how the presence of salt affects the performance of the chemical and physical processes. Specifically, the processes covered will be coagulation-flocculation, sedimentation, dehydration and disinfection since they are commonly applied to both urban and industrial wastewater. Since the interaction between particles of the wastewater itself or of the biomass generated during the biological treatment is important in the first three processes, the approach of the DLVO theory, established by Derjaguin, Landau, Verwey, and Overbeek in the 1940s will be briefly discussed below (Derjaguin & Landau, 1941; Verwey & Overbeek, 1948).

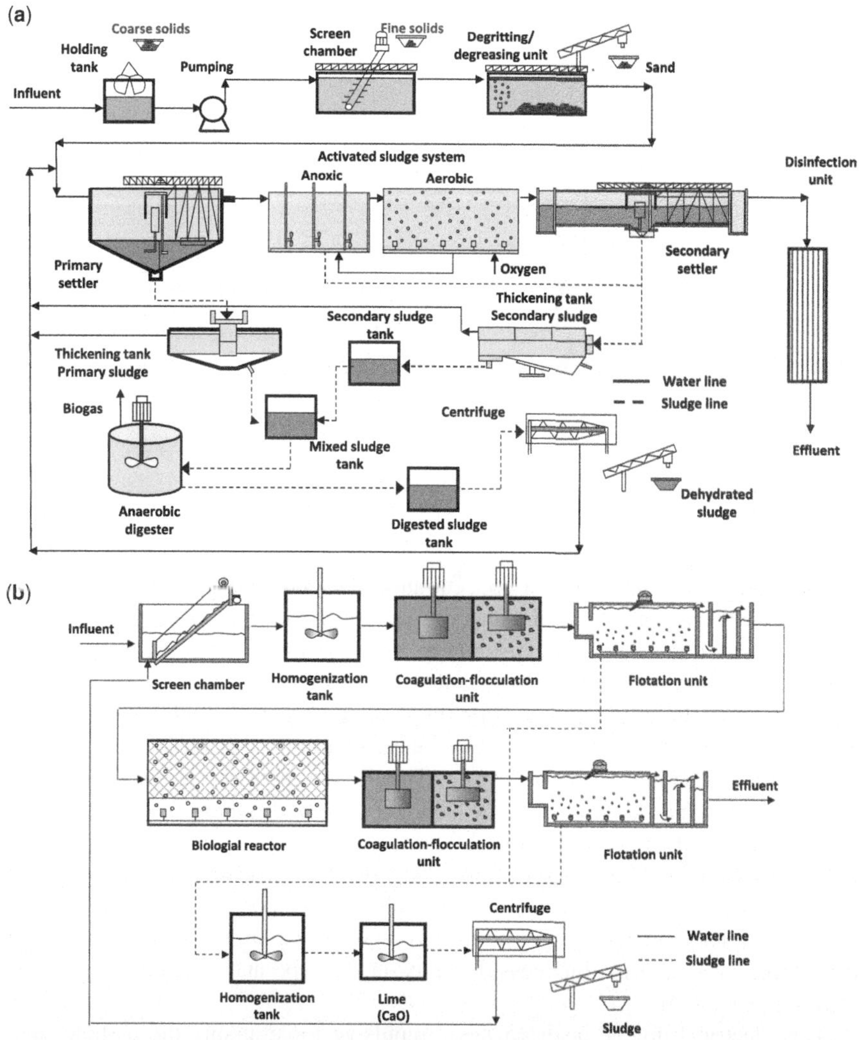

Figure 2.2 Layout of wastewater treatment plants (WWTPs) for treating: (a) urban wastewater and (b) wastewater from the food industry.

Most sludge particles in nature are negatively charged which causes some positive ions to form a rigid adjacent layer around their surface. This counter ion layer is known as the Stern layer (Figure 2.3). Other additional positive ions are still attracted to the negative colloid, but these are now repelled by the Stern layer, as well as by other positive ions that try to approach the colloid. This dynamic equilibrium results in the formation of a diffuse layer of counter ions.

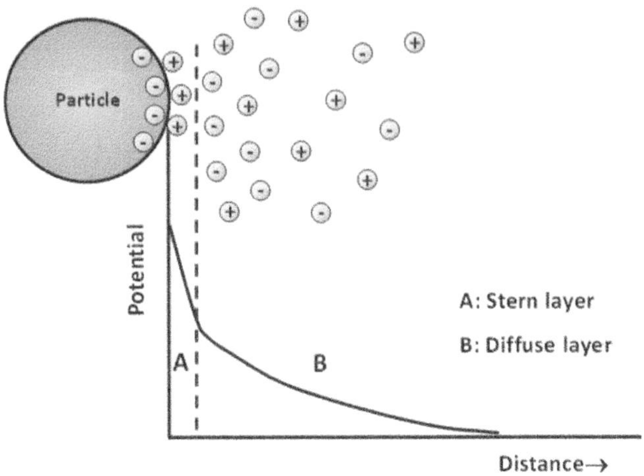

Figure 2.3 Diagram of electrical double layer. DLVO theory.

These counter ions have a high concentration near the surface, which gradually decreases with distance, until a balance is achieved with the concentration of the counter ions within the solution. Similarly, although opposite, in the diffuse layer there is a deficit of negative ions, called co-ions because they have the same charge as the colloid. Their concentration gradually increases with the distance to the colloid, while the repulsive forces of the colloid are compensated by positive ions, until equilibrium is reached again. The diffuse layer can be visualized as a charged atmosphere surrounding the colloid. At any distance from the surface, the charge density is equal to the difference in concentration between positive and negative ions. The charge density is much higher near the colloid and gradually decreases to zero when the concentrations of positive and negative ions are similar. The counter ions of both Stern and diffuse layers are what is called the double layer and its thickness depends on the type and concentration of the ions in the solution.

This electrical double layer creates a repulsive force among the particles and prevents them from agglomerating into large flocs (Figure 2.4). However, colloidal particles in a suspension also experience attractive forces as a result of dipole-dipole interactions also generally known as van der Waals interactions. According to the DLVO theory, the electrostatic double layer forces and the van der Waals forces are independent and, therefore, can be superimposed or added at each interacting distance between two particles. The van der Waals attraction force is an inverse power law function of the separation distance and is insensitive to the ionic strength of the bulk liquid. The electrostatic repulsion force decays exponentially with separation distance and depends on the ionic strength of the bulk liquid since it controls the range of the double layer

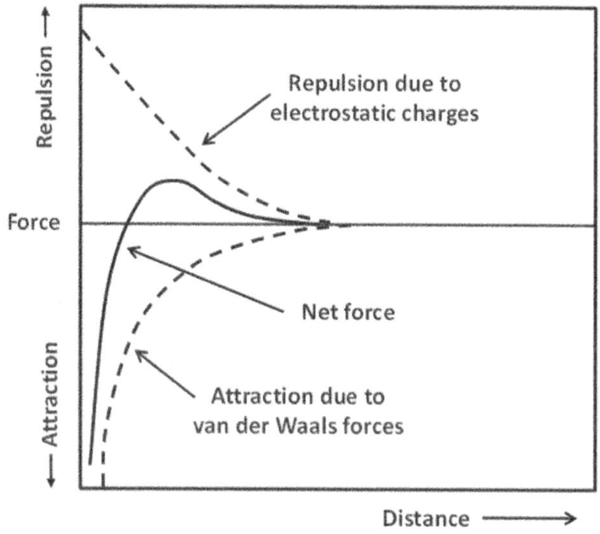

Figure 2.4 Forces involved between two colloidal particles in a suspension.

interaction. The sum of both forces gives the net interaction energy at a particular separation distance. This net interaction energy is always attractive at short distances and repulsive at intermediate distances while, at large distances, it can be attractive or repulsive depending on the ionic strength of the bulk liquid. In any case, the net interaction energy always presents a maximum of repulsion, called the energy barrier, which should be overcome by the kinetic energy of the particles for them to approach one another. In order to promote particle agglomeration, the energy barrier can be decreased by increasing the ionic strength of the bulk liquid or by adding active compounds to directly affect the particle charge. This can be done in practice by increasing the salinity of the medium.

2.2 COAGULATION-FLOCCULATION

The main purpose of coagulation is to overcome the electrostatic repulsion between colloids caused by the negative electrical double layer which surrounds them in water and maintains their stability by preventing them from adhering to each other and forming larger particles. Particle destabilization can occur through four mechanisms: double layer compression, charge neutralization, colloid entrapment and intraparticle bridging (Teh *et al.*, 2016). The double layer compression mechanism involves the reduction in this layer around the colloidal particles, by the increase of ionic strength via salts addition, which allows them to approach more closely before experiencing repulsion and, then, leads to their coagulation.

This mechanism alone is not feasible to treat wastewater at an industrial scale due to the massive amount of salt needed but double layer compression improves the efficiency of the coagulation process when it occurs by means of charge neutralization or intraparticle bridging mechanisms.

Flocculation can occur through charge neutralization by the adsorption of positively charged dissolved metal hydroxide species (Figure 2.5). In this mechanism, the charge neutralization cancels the electrostatic repulsion between colloids, and they can link to form flocs which can be easily removed from wastewater. Nevertheless, an excessive addition of metal salts to wastewater causes colloids to acquire a positive surface charge which avoids flocs formation (restabilization). Metal salts (containing iron or aluminium) are generally added to wastewater to effectively achieve the neutralization of the negative surface charge of many types of particles. This mechanism is promoted by the presence of salts since the high ionic strength of salt water destabilizes colloids at a wider range of pH and coagulant dose conditions (Edzwald *et al.*, 1974; Hilal *et al.*, 2008; Schneider *et al.*, 2013; Younker & Walsh, 2014; Zhang *et al.*, 2014). This promotion is due to the double layer compression which rapidly becomes smaller as the ionic strength increases (Hilal *et al.*, 2008). However, the presence of salts is beneficial up to a certain concentrations beyond which the rise of salt concentration hinders coagulation (Wang *et al.*, 2013). This fact is related to the adsorption of Ca^{+2} or Na^{+} on the surface of the particles which increases their positive surface charge and stabilizes them. This absorption phenomenon occurs at high salt concentrations and causes deviations from results expected by the DLVO theory (Pashley & Israelachvili, 1984).

Aluminium and iron salts can be added to wastewater at a dosage higher than the solubility of their amorphous hydroxides. The metal hydroxides precipitation and their later settling entrap and enmesh particles and colloids and remove them from the wastewater. Salt presence has a negative effect on the efficiency of this

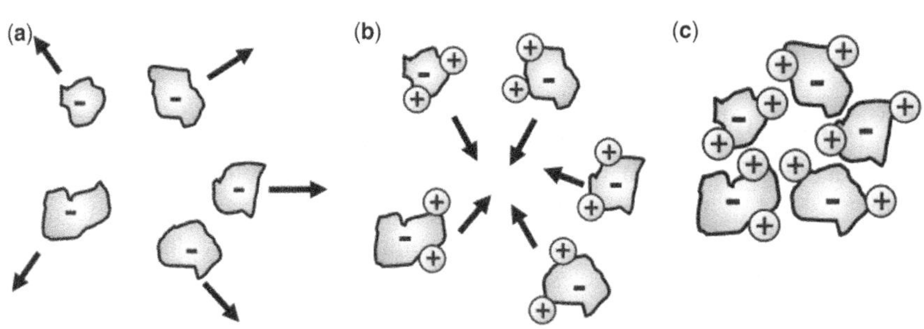

Figure 2.5 Particle destabilization through charge neutralization: (a) Repulsion forces maintain stability of colloidal particles; (b) Positively charged dissolved metal hydroxide species neutralize particle charges and van der Waals forces predominate; (c) Particles form flocs and settle.

mechanism since the effective concentration of the metal ions would decrease with increasing ionic strength because their activity decreases (Wang *et al.*, 2010).

The fourth mechanism of coagulation is interparticle bridging, usually achieved by an addition of water-soluble polymers. This mechanism requires a region of the polymer chain to attach to the particle surface while the free part of the polymer chain extends to the surrounding bulk liquid for adherence with other particles. According to literature, coagulation by polymers addition is enhanced under saline conditions (Ajao *et al.*, 2018; Hilal *et al.*, 2008; Ji *et al.*, 2013). Hilal *et al.* (2008) attribute this positive effect to the double layer compression of both particles and polymer molecules. Moreover, Ji *et al.* (2013) found that, in a solution of high salinity, the polymers have a more condensed conformation which enhances their adsorption density (amount of adsorbate adsorbed per unit area).

2.3 SETTLING

Settling problems of activated sludge flocs in saline wastewaters are often reported in the literature. Several factors have been provided to explain this phenomenon (Woolard & Irvine, 1995): (1) the buoyancy force raises with increasing wastewater salt concentration due to the increase in the water density which reduces the terminal settling velocity according to Stokes law; (2) high salt concentrations cause disintegration of activated sludge flocs and cell plasmolysis due to the increase of osmotic pressure, which results in a reduction in particle size and density and, therefore, the formation of 'pin-point' sludge flocs (Kargi & Dincer, 1997; Mesquita *et al.*, 2009; Moon *et al.*, 2003); (3) the filamentous bacteria which provide the mechanical integrity and structure of the flocs are seriously affected by high salt concentrations (Zhan, 1999); and (4) saline environments cause the reduction of protozoans in the secondary sludge and the effluent turbidity increases since these microorganisms are predators of dispersive microorganisms which are not settleable in wastewater (Salvadó *et al.*, 2001). In contrast, the sludge volume index (SVI) of biomass tends to diminish with the increase of the salt content (Campos *et al.*, 2002; Mesquita *et al.*, 2009; Wu *et al.*, 2008). The SVI decrease is probably due to the increase hydrophobicity (Zhan, 1999) or the morphological change of activated sludge flocs (Ng *et al.*, 2005). In spite of the SVI value decreasing at high salt content, the quality of the settled effluent worsens in terms of turbidity indicating a disaggregation phenomenon, given the intrinsic low mechanical properties of pin-point flocs, and the formation of new ones by deflocculation of larger aggregates (Mesquita *et al.*, 2009; Wu *et al.*, 2008). Ng *et al.* (2005) explained that SVI and turbidity are not necessarily related directly because they reflect different physical issues: the SVI reflects the compactability of the biomass whereas turbidity reflects the ability of dispersed organisms to flocculate. In fact, the increase of biomass compaction could be explained by the floc size shrinking in saline water (Chu *et al.*, 1997) while the loss of flocculation ability could be attributed to the destabilization of the floc by an ion exchange mechanism with Na^+ ions (Gordon & Millero, 1984; Higgins & Novak, 1997) or to enhancement of hydrophobic interactions and weakening of electrostatic interactions due to the salt presence (Lindahl *et al.*, 1981).

Aerobic granular sludge technology has been extensively studied to improve WWTPs because aerobic granules can settle faster than flocs due to their compact structure, but their settling properties are also affected by the presence of salts. Winkler *et al.* (2012) studied the effect of salinity on the granular biomass settling velocity in a range of 0 to 40 g NaCl/L and found that the settling velocity linearly decreased with salinity. These authors attribute this effect to the increase of water density since the density of granules is only slightly different from water (1005–1070 kg/m^3 (Bassin *et al.*, 2012; Batstone & Keller, 2001; Etterer & Wilderer, 2001)) and, therefore, changes in the density of the water have a significant impact on the settling behaviour of granules. To carry out their experiments, Winkler *et al.* (2012) pre-incubated aerobic granules in salt solutions because they tended to float if they were not pre-incubated. This fact would suggest that a sudden increase of the bulk liquid salinity, as happens in some industrial effluents, could cause biomass wash-out.

In the case of inorganic solids, most of the works studying the salt presence effect on their settling properties were not focused on wastewater treatment but on the behaviour of estuary sediments. These works show that the settling velocity of inorganic particles improves when salt concentration increases due to the flocculation of the fine-grained sediments in suspension (Ou *et al.*, 2016; Portela *et al.*, 2013). However, the response of the settling velocity to salinity does not have a monotone trend, since there is a salinity range where the settling velocity is maximum (Xu *et al.*, 2018). This observation disagrees with that expected by the DLVO theory since according to this theory, at high electrolyte concentrations, the net interaction between two surfaces would be attractive at all separation distances and would result in a strong adhesion. To explain this fact, Gordon and Millero (1984) hypothesize that there is a salt concentration where the theoretical double-layer thickness is small and relatively constant and where an ion-exchange mechanism starts to play an important role.

2.4 DEWATERING

Dewatering is a physical process commonly implemented in the sludge line of WWTPs which aims to reduce the sludge water content, and therefore its volume, to decrease the transportation costs, since they can represent up to 50% of the WWTP operating costs (Collivignarelli *et al.*, 2019). The dewatering technique to be applied must be chosen taking into account the amount and characteristics of the sludge generated and the requirements of its final destination.

Water contained in sludge can be classified in four forms according to the degree of removal difficulty: free, interstitial, surface and intracellular (Vaxelaire & Cézac,

2004) (Figure 2.6). The term free water is related to that water located in voids and it is not associated with solids (not affected by capillary force). It accounts for approximately 70% of the total water (Chen *et al.*, 2015a). The interstitial water is bound by capillary forces between the sludge flocs (Mikkelsen & Keiding, 2002) while surface water is water adsorbed or adhered onto the surface of particles. Finally, intracellular water is internal water which is tightly bound to the solids (Vaxelaire & Cézac, 2004). The sum of the three latter forms is known as bound water (Lee & Hsu, 1995). Free water can be separated from sludge by gravity while chemical conditioning prior to the use of mechanical devices is required to remove interstitial and surface water. For intracellular water removal, the structure containing it must be broken and this can be done through heat treatment.

In WWTPs, the majority of the sludge comes from biological treatment and it is known as waste activated sludge (WAS) which is composed of microorganisms, extracellular polymeric substances (EPS), organic debris, and inorganic colloidal particles (Christensen *et al.*, 2015). The sludge dewaterability is hindered by the EPS fraction, which is negatively charged and highly hydrated and captures a large amount of water (He *et al.*, 2016).

In general, the mechanical dewatering devices, such as the belt filter press, vacuum filter, and centrifuge, are commonly employed for decreasing the sludge moisture content (Bennamoun *et al.*, 2013). Their efficiency depends mainly on the sludge dewaterability which is usually evaluated by two parameters: capillary

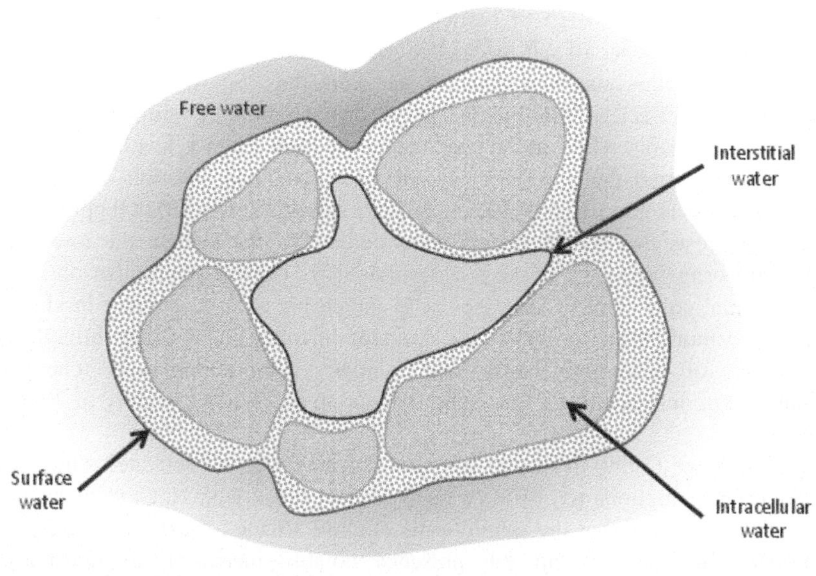

Figure 2.6 Different types of water contained in sludge.

suction time (CST) and specific resistance in filtration (SRF) (To *et al.*, 2016). It has been found that the floc size and density are the major factors affecting dewaterability. Sludge with higher floc density and larger floc size is much easier to dewater and settle (Wen *et al.*, 1997). In this sense, the application of the coagulation and flocculation processes before or during mechanical dewatering promotes particle agglomeration and, therefore, increases floc size and density. As coagulation can disrupt the electrical double layer of the particles, the coagulated sludge becomes tightly packed and is able to release more bound water from the sludge to the bulk solution. Although the coagulation/flocculation mechanisms in sludge dewatering are similar to those in pollutants removal, the treatment conditions and objectives of the two processes are fundamentally different. The latter focuses on the aggregation and separation of pollutants promoted by coagulants/flocculants; in addition to this effect, the former one still emphasizes the formation of a suitable structure of sludge cake with good filtration performance (Raynaud *et al.*, 2012).

Salinity is also believed to have a significant effect on sludge dewaterability (Chu *et al.*, 1997; Cousin & Ganczarczyk, 1998) and this topic has recently been gaining importance in urban WWTPs because of the use of seawater in municipal flushing systems in coastal areas (Wang *et al.*, 2009). Lo *et al.* (2001) measured the SRF of sludge under three different salinity levels (5 000, 10 000 and 20 000 ppm) and found that the sludge SRF decreases with increasing salinity (a reduction in SRF value of approximately 30% is achieved by doubling the salinity level). This result agrees with that expected according to the DLVO theory since an increase of the salt concentration causes the compression of the double layer thickness, reducing the surface charge of particles and promoting particle agglomeration. Moreover, the presence of salt in the sludge might also cause a salting out effect, which enhances the release of bound water from the flocs to the bulk solution. Therefore, the effect of salinity on sludge dewaterability is similar to that of chemical coagulants when the sludge reaches a high salinity level. In fact, Lo *et al.* (2001) found that the effect of chemical coagulant addition on the SRF value obtained is almost insignificant at high salinity levels (20 000 ppm). This finding suggests that the higher the salts content of the sludge, the lower the chemical dosage required to meet a desired SRF. These authors also observed that the final solid content increases with increasing salinity content in sludge, since the osmotic effect promotes the drainage of intracellular water through the membrane of microorganisms, and sludge without previous chemical conditioning could achieve a final solid content up to 27% at a salinity of 20 000 ppm.

Jean and Lee (1999) evaluated filtration characteristics of activated sludge for different concentrations (0–4% w/w) and contact times with NaCl (1 h, 1 d and 1 w). They observed that the salinity only slightly affected the dewaterability for a contact time of 1 h, but salt presence notably increased the dewatering efficiency for contact times longer than 1 day. They also found that the dewatering efficiency achieved a maximum value at a NaCl concentration of 2% and then it decreased. This finding is related to the change of the bacteria aggregation mechanism or of sludge flocs stability at a critical salinity concentration (Gordon & Millero, 1984; Higgins & Novak, 1997). Below this critical concentration, the interaction among bacteria is favoured by the double layer compression, according to DLVO theory (Zita & Hermansson, 1994). At a higher concentration, the tendency for bacteria aggregation decreases with the salinity level since Na^{+} can replace the bridging capability of Ca^{+2} in the sludge floc. Therefore, the sludge flocs would be destabilized at high salts concentrations due an ion exchange mechanism (Bruus *et al.*, 1992; Rasmussen *et al.*, 1994). This breakdown of aggregates in the compacted filter cake could markedly increase the collapse of the cake structure, leading to a worse dewaterability. This ion exchange was confirmed by Jean and Lee (1999) by comparing the amount of Ca^{+2} initially present in the sludge to that of sludge exposed to high salt concentrations for a contact period of 1 day. However, Zita and Hermansson (1994) did not observe the release of Ca^{+2} after exposing sludge to high salt concentrations and, then, proposed that the conformation change of extracellular polymer could be involved in destroying sludge flocs.

2.5 FLOTATION

There are different types of flotation processes, with dispersed/induced and dissolved air flotation being the most commonly applied for wastewater treatment (Prakash *et al.*, 2018; Rubio *et al.*, 2002). Dispersed air flotation operates by generating gas bubbles, using spargers or gas distributors located at the bottom of a tank, which are dispersed into the bulk liquid. The range of bubble size obtained is about 700–1500 μm in diameter (Rubio *et al.*, 2002). This technique is widely used to treat industrial wastewater from refineries, natural gas processing plants, and petrochemical and chemical plants (El-Kayar *et al.*, 1993). Dissolved air flotation involves supersaturating water with air under high pressure, which generates microbubbles when released into water under atmospheric pressure. The produced microbubbles size is between 20 and 100 μm in diameter (Edzwald, 2010). Dissolved air flotation is used to separate water pollutants such as particles or microorganisms in a size range of 10 to 100 μm (Aliff Radzuan *et al.*, 2016) or grease from urban wastewater. To enhance the flotation process, chemical coagulants are often added in order to increase the flocculent structure of floated particles so they can be easily entrapped in the air bubbles (Tetteh & Rathilal, 2019).

Salt presence affects both particle and bubble properties and its overall effect is the promotion of the flotation which implies saving operating costs since smaller amounts of chemical coagulants are required (Quinn *et al.*, 2007). This enhancement is attributed to: (1) the reduction in the electrostatic interactions between particle and bubble that increases particle-bubble attachment efficiencies

(Hewitt *et al.*, 1994; Pugh *et al.*, 1997); and (2) the inhibition of bubble coalescence which favours the formation of small stable bubbles, increasing the particle-bubble collision probability (Bournival *et al.*, 2012; Pugh *et al.*, 1997). Nevertheless, there is no consensus on what the mechanisms are that cause this improvement (Ofori *et al.*, 2005; Wang and Peng, 2013; Wang and Peng, 2014). Taking into account the contact angle measured for coal particles in NaCl solutions, Klassen and Mokrousov (1963) attributed this improvement to the lowering of the electrochemical potential of particle surfaces, due to adsorbed ions reducing the stability of the hydration layers surrounding the particles increasing their surface hydrophobicity. A similar hypothesis was proposed by Arnold and Aplan (1986). Nevertheless, Kurniawan *et al.* (2011) think that the compression of the electrical double layer by electrolytes is the main reason for this improvement since it causes the reduction of electrical repulsion between particles and bubbles and, subsequently, facilitates their attachment process.

Bubble coalescence is the process where two or more bubbles come together to generate a new one of larger volume. This occurs in three steps: collision, thinning of the liquid film between bubbles (drainage) and rupture (coalescence) (Figure 2.7a). Firstly, the bubbles come into contact in the liquid phase and the contact surfaces flatten and trap a liquid film between them whose initial thickness is 10^{-3}–10^{-4} cm. Then, after a certain contact time, the film is thinned to approximately 10^{-6} cm and, finally, it breaks and generates a new bubble (coalescence). This last step is fast compared to the other two. The presence of salts can inhibit bubble coalescence by retarding the thinning of the liquid film

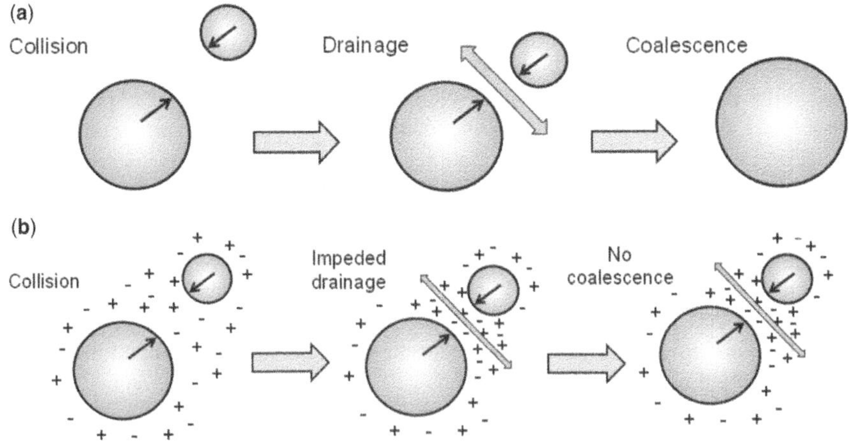

Figure 2.7 Behaviour of air bubbles in water: (a) bubble coalescence when electrolytes are not present; (b) bubble coalescence inhibition at high electrolyte concentrations.

between bubbles, because they cannot diffuse away quickly, raising the local salt concentration and therefore the surface tension. The surface tension increase prevents draining of the liquid film and retards coalescence (Marrucci, 1969). Moreover, electric charge that accumulates on the surfaces of bubbles produces a repulsive force, preventing them from approaching each other (Figure 2.7b) (Cui *et al.*, 2003; Marrucci & Nicodemo, 1967). In fact, Craig *et al.* (1993) suggest that the coalescence in pure water might be caused by a strong hydrophobic attractive force and the adsorption of certain ion pairs at the air/water interface reduces it.

2.6 PRECIPITATION

Salt presence can significantly affect the chemical equilibria in the bulk liquid, such as the solubility product, and, therefore, the efficiency of chemical precipitation processes. In non-electrolyte solutions, the intermolecular forces are mainly due to weak van der Waals interactions, but in ionic solutions electrostatic interactions of solute-solvent and solute-solute molecules, governed by Coulomb's law, are significant. Consequently, the behaviour of an electrolyte solution deviates considerably from that of an ideal solution. If ions of soluble salts are present in a solution containing ions of sparingly soluble salts, the formers will be attracted to those slightly soluble salt ions of opposite charge and will tend to surround them. This phenomenon creates an ionic atmosphere that reduces the attraction between cations and anions of the sparingly soluble salts and, therefore, results in a low tendency for them to meet, react and precipitate. That means the 'effective' concentration of these cations and anions is below the actual concentration and, for this reason, the activity of the individual components and not the concentration itself should be used to calculate chemical equilibria under the influence of ionic strength (Willey, 2004). For this purpose, activity coefficients (f) can be calculated using the extended Debye-Hückel equation (2.1):

$$-\log(f) = \frac{Az^2\sqrt{I}}{1 + ba\sqrt{I}} \tag{2.1}$$

where A and b are empirical parameters whose values, at 25°C, are 0.51 and 0.33, respectively; z is the ion charge; a is the hydrated radius of the ion (nm); and I is the ionic strength (Molar) which can be calculated by Equation (2.2):

$$I = \frac{1}{2}\sum_{i=1}^{n} c_i z_i^2 \tag{2.2}$$

where c is the ion concentration (Molar).

According to the extended Debye-Hückel equation, the activity coefficient value asymptotically declines from a value of 1 at zero ionic strength to a low value at high ionic strength. However, the experimental data show that most ions reach a minimum activity coefficient below an ionic strength of 1 M and then, their activity coefficient increases. This fact can be explained because, at high ionic strength, their own electrostatic interactions, which limit ion mobility at low ionic strength, impose a pairing of charges that increases the effective activity. For this reason, the application of the extended Debye-Hückel equation to ionic strength values below 0.1 M is advisable. For calculating the activity coefficients at high values of ionic strength, the Pitzer equations provide a reasonable accuracy (Sheikholeslami & Ong, 2003).

In some cases, there are specific interactions between ions and solutes come close enough to get in direct contact and are considered new species. These new species are called ion pairs, when ions are separated by water molecules but share their first hydration shell, or complexes, when ions are in contact and share electrons. Ions pairs and complexes present low reactivity, so their formation lowers the effective concentration of solutes and, therefore, increases their solubility. This is the case in heavy metals removal under saline conditions where the formation of complexes between chloride and the heavy metals stabilizes them avoiding their precipitation (Lv *et al.*, 2018).

In practical terms, since the actual solubility of ionic compounds under saline conditions is higher than that determined in pure water, to remove ions by chemical precipitation in the presence of salts would imply an increase of the reagent addition requirements.

Ionic salts also affect the precipitation rate of sparingly soluble salts and the size distribution and morphology of the formed crystals due to their influence on the water of solvation and bulk solvent structure (Kowacz *et al. et al.*, 2010; Zhang *et al.*, 2019). However, the effects observed vary depending on the studied salt and on the ions present in the bulk liquid (Hu *et al.*, 2014; Sheikholeslami & Ong, 2003; Tai & Chen, 1998; Zhang *et al.*, 2019; Zuddas & Mucci, 1998).

2.7 DISINFECTION

The effect of salinity on the disinfection unit depends on the technology applied to remove microorganisms. One of the most commonly applied technologies to carry out disinfection is based on the radiation of UV light with wavelengths between 100 and 400 nm since it causes damage to the DNA of the pathogenic organism. This damage is mainly due to cyclobutane pyrimidine dimers (CPDs) which are photoproducts generated from irradiated DNA. CPDs can hinder both replication of DNA and transcription of DNA into RNA, avoiding microorganism reproduction and protein synthesis, respectively (Harm, 1980).

The effect of salinity on UV disinfection is not clear since literature reports contradictory results. Chen *et al.* (2018) studied the effect of salinity on the log removal and inactivation rate of *V. cholerae* using medium and low pressure (MP and LP) UV lamps and found that a salinity level of 1% or higher had a negative impact on the performance of both types of lamps especially at high UV doses (3 mJ/cm^2 for MP and 4 mJ/cm^2 for LP) in spite of this, no significant effects of salinity were found on the CPDs levels. These authors had previously found similar results when studying the inactivation of *Enterococcus faecalis* (Chen *et al.*, 2015b). Rubio *et al.* (2013) evaluated the disinfection efficiency of *E. coli* by UV radiation and they also observed that the UV inactivation rate decreased when increasing the salt concentration. These results could be explained by clustering or adsorption of the microorganisms to surfaces promoted at high ionic strength values which cause a shield effect against UV radiation. Contrary to this, Shang *et al.* (2009) found that higher salinity resulted in higher level of inactivation of faecal coliform bacteria at a UV-C dose of 12 mJ/cm^2. This could be related to the formation of radicals that interact with bacteria (Buschmann *et al.*, 2005; Legrini *et al.*, 1993). It is known that inorganic ions (Cl^-, HCO_3^- and SO_4^{-2}) exposed to UV light can absorb it and form radicals that interact with bacteria (Buschmann *et al.* 2005; Rincon and Pulgarin, 2004; Wright & Cairns 1998). Hence, the overall effect of salinity on UV inactivation depends on both UV light attenuation and radical formation.

Advanced oxidation processes (AOPs) such as UV/H_2O_2 and UV/TiO_2, which use powerful oxidizing radicals (mostly hydroxyl (·OH)) that instantaneously react with microorganisms, achieve disinfection efficiencies higher than those of UV alone in water with high salt concentrations. Nevertheless, in the case of the UV/TiO_2 treatment, an important negative effect of salinity on the inactivation rates of *Enterococcus faecalis* was found due to the ion-blockage of the active sites on the photocatalytic surface (Moreno-Andrés *et al.*, 2017). Rubio *et al.* (2013) also found that photo-Fenton and H_2O_2/UV treatments were effective to deactivate *E. coli* in a wide range of salinity levels but they were affected by the presence of bicarbonate ions since they were the main scavenger of hydroxyl radicals generated during these treatments.

UV-inactivated microorganisms regain activity through the repair of pyrimidine dimers in the genome by utilizing the energy of UV-A and visible light (320–500 nm) with a specific enzyme, photolyase. This phenomenon is called photoreactivation and may be of significant concern especially when UV-treated wastewater is discharged to a natural water environment, because UV-inactivated microorganisms are exposed to sunlight which includes UV-A and visible light. Oguma *et al.* (2013) observed that photoreactivation of *E. coli* was significantly suppressed in NaCl solutions at concentrations of 2.4% or higher, but not in NaCl solutions at 1.9% or lower. Therefore, *E. coli* in UV-treated wastewater may possess the potential to perform photoreactivation after being discharged to water bodies where salinity is rather low. In this sense, Shang *et al.* (2009) proposed to carry out disinfection by UV radiation in the presence of TiO_2 because, under these conditions, the repression of bacteria photoreactivation persisted, independent of the salinity level. In the case of salt tolerant organisms, such as *Enterococcus faecalis*, salinity has a beneficial effect on photoreactivation because high salt levels may cause the expression of CPD photolyase leading to DNA double-strand breaks (Chen *et al.*, 2015b; Cheng *et al.*, 2007).

2.8 CONCLUSIONS

The presence of salt in wastewater has a positive effect on the efficiency of coagulation, settling (of inorganic particles), dewatering and flotation processes because it promotes aggregation of particles. The efficiency of these processes improves up to a critical salt concentration beyond which a subsequent increase in salinity causes a worsening of the particles' aggregation. In addition, in the case of flotation, the presence of salt inhibits the bubble coalescence, which enhances its efficiency. Salinity has a negative effect on the settling process of biological sludge since salts presence decreases the growth rate of essential microorganisms both to maintain the floc structure and to obtain a good effluent quality in terms of solids. Salinity also negatively affects the removal efficiency of the chemical precipitation process due to the increase in the solubility of ionic compounds under saline conditions. In the case of the UV disinfection process, the effect of salinity is not clear due to it having different effects on several of the mechanisms involved.

REFERENCES

Ajao V., Bruning H., Rijnaarts H. and Temmink H. (2018). Natural flocculants from fresh and saline wastewater: Comparative properties and flocculation performances. *Chemical Engineering Journal*, **349**, 622–632.

Aliff Radzuan M. R., Abia-Biteo Belope M. A. and Thorpe R. B. (2016). Removal of fine oil droplets from oil-in-water mixtures by dissolved air flotation. *Chemical Engineering Research and Design*, **115**, 19–33.

Arnold B. J. and Aplan F. F. (1986). The effect of clay slimes on coal flotation, Part II: the role of water quality. *International Journal of Mineral Processing*, **17**, 243–260.

Bassin J. P., Winkler M. K. H., Kleerebezem R., Dezotti M. and van Loosdrecht M. C. M. (2012). Improved phosphate removal by selective sludge discharge in aerobic granular sludge reactors. *Biotechnology Bioengineering*, **109**, 1919–1928.

Batstone D. J. and Keller J. (2001). Variation of bulk properties of anaerobic granules with wastewater type. *Water Research*, **35**, 1723–1729.

Bennamoun L., Arlabosse P. and Leonard A. (2013). Review on fundamental aspect of application of drying process to wastewater sludge. *Renewable & Sustainable Energy Reviews*, **28**, 29–43.

Bournival G., Pugh R. J. and Ata S. (2012). Examination of NaCl and MIBC as bubble coalescence inhibitor in relation to froth flotation. *Minerals Engineering*, **25**, 47–53.

Bruus J. H., Nielsen P. H. and Keiding K. (1992). On the stability of activated sludge flocs with implications to dewatering. *Water Research*, **26,** 1597–1604.

Buschmann J., Canonica S., Lindauer U., Hug S. J. and Sigg L. (2005). Photoirradiation of dissolved humic acid induces arsenic (III) oxidation. *Environmental Science and Technology*, **39**, 9541–9546.

Campos J. L., Mosquera-Corral A., Sánchez M., Méndez R. and Lema J. M. (2002). Nitrification in saline wastewater with high ammonia concentration in an activated sludge unit. *Water Research*, **36**, 2555–2560.

Chen Z., Zhang W. J., Wang D. S., Ma T. and Bai R. Y. (2015a). Enhancement of activated sludge dewatering performance by combined composite enzymatic lysis and chemical re-flocculation with inorganic coagulants: kinetics of enzymatic reaction and re-flocculation morphology. *Water Research*, **83**, 367–376.

Chen P. Y., Chu X. N., Liu L. and Hu J. Y. (2015b). Effects of salinity and temperature on inactivation and repair potential of *Enterococcus faecalis* following medium- and low-pressure ultraviolet irradiation. *Journal of Applied Microbiology*, **120**, 816–825.

Chen P. Y., Chu X. N., Liu L. and Hu J. Y. (2018). Effect of salinity on medium- and low-pressure UV disinfection of *Vibrio cholera*. *Water Science and Technology*, **77**, 655–661.

Cheng L., Qiao D. R., Lu X. Y., Xiong Y., Bai L. H., Xu H., Yang Y. and Cao Y. (2007). Identification and expression of the gene product encoding a CPD photolyase from Dunaliella salina. *Journal of Photochemistry and Photobiology A: Chemistry*, **B87**, 137–143.

Christensen M. L., Keiding K., Nielsen P. H. and Jorgensen M. K. (2015). Dewatering in biological wastewater treatment: a review. *Water Research*, **82**, 14–24.

Chu C. P., Feng W. H., Tsai Y. H. and Lee D. J. (1997). Unidirectional freezing of waste-activated sludge: the presence of sodium chloride. *Environmental Science and Technology*, **31**, 1512–1517.

Collivignarelli M. C., Abbà A., Miino M. C. and Torretta V. (2019). What advanced treatments can be used to minimize the production of sewage sludge in WWTPs? *Applied Sciences*, **9**, 2650.

Cousin C. P. and Ganczarczyk J. J. (1998). Effects of salinity on physical characteristics of activated sludge flocs. *Water Quality Research Journal of Canada*, **33**, 565–587.

Craig V. S. J., Ninham B. W. and Pashley R. M. (1993). Effect of electrolytes on bubble coalescence. *Nature*, **364**, 317–319.

Cui Q., Chandra S. and McCahan S. (2003). The effect of dissolving salts in water sprays used for quenching a hot surface: Part 1-Boiling of single droplets. *Journal of Heat Transfer-ASME Journal*, **125**, 326–332.

Derjaguin B. and Landau L. (1941). Theory of the stability of strongly charged lyophobic sols and of the adhesion of strongly charged particles in solutions of electrolytes. *Acta Physicochimica URSS*, **14**, 633–662.

Edzwald J. K. (2010). Dissolved air flotation and me. *Water Research*, **44**, 2077–2106.

Edzwald J. K., Upchurch J. B. and O'Melia C. R. (1974). Coagulation in estuaries. *Environmental Science and Technoogy*, **8**, 58–62.

El-Kayar A., Hussein M., Zatout A. A., Hosny A. Y. and Amer A. A. (1993). Removal of oil from stable oil-water emulsion by induced air flotation technique. *Separations Technology*, **3**, 25–31.

Etterer T. and Wilderer P. A. (2001) Generation and properties of aerobic granular sludge. *Water Science and Technology*, **43**, 19–26.

Gordon A. S. and Millero F. J. (1984). Electrolyte effects on attachment of an estuarine bacterium. *Applied and Environmental Microbiology*, **47**, 495–499.

Harm W. (1980). Biological Effects of Ultraviolet Radiation. Cambridge University Press, New York, NY, *USA*.

He D. Q., Luo H. W., Huang B. C., Qian C. and Yu H. Q. (2016). Enhanced dewatering of excess activated sludge through decomposing its extracellular polymeric substances by a Fe@Fe_2O_3-based composite conditioner. *Bioresource Technology*, **218**, 526–532.

Hewitt D., Fornasiero D. and Ralston J. (1994). Bubble particle attachment efficiency. *Minerals Engineering*, **7**, 657–665.

Higgins M. J. and Novak J. T. (1997). The effect of cations on the settling and dewatering of activated sludges: laboratory results. *Water Environment Research*, **69**, 215–224.

Hilal N., Al-Abri M., Moran A. and Al-Hinai H. (2008). Effects of heavy metals and polyelectrolytes in humic substance coagulation under saline conditions. *Desalination*, **220**, 85–95.

Hu Y. B., Wolf-Gladrow D. A., Dieckmann G. S., Völker C. and Nehrke G. (2014). A laboratory study of ikaite ($CaCO_3{\cdot}6H_2O$) precipitation as a function of pH, salinity, temperature and phosphate concentration. *Marine Chemistry*, **162**, 10–18.

Jean D. S. and Lee D. J. (1999). Effects of salinity on expression dewatering of waste activated sludge. *Journal of Colloid and Interface Science*, **215**, 443–445.

Ji Y., Lu Q., Liu Q. and Zeng H. (2013). Effect of solution salinity on settling of mineral tailings by polymer flocculants. *Colloids and Surfaces A: Physicochemical and Engineering Aspects*, **430**, 29–38.

Kargi F. and Dincer A. R. (1997). Biological treatment of saline wastewater by fed-batch operation. *Journal of Chemical Technology and Biotechnology*, **69**, 167–172.

Klassen V. I. and Mokrousov V. A. (1963). An introduction to the theory of flotation, second ed. Butterworths, London.

Kowacz M., Prieto M. and Putnis A. (2010). Kinetics of crystal nucleation in ionic solutions: Electrostatics and hydration forces. *Geochimica et Cosmochimica Acta*, **74**, 469–481.

Kurniawan A. U., Ozdemir O., Nguyen A. V., Ofori P. and Firth B. (2011). Flotation of coal particles in $MgCl_2$, NaCl, and $NaClO_3$ solutions in the absence and presence of Dowfroth 250. *International Journal of Mineral Processing*, **98**, 137–144.

Lee D. J. and Hsu Y. H. (1995). Measurement of bound water in sludges: a comparative study. *Water Environment Research*, **67**, 310–317.

Legrini O., Oliveros E. and Braun A. M. (1993). Photochemical processes for water treatment. *Chemical Reviews*, **93**, 671–698.

Lindahl M., Fads A., Wadstrom T. and Hjerten S. (1981). A new test based on 'salting out' to measure relative surface hydrophobicity of bacterial cells. *Biochimica et Biophysica Acta*, **677**, 471–476.

Lo I. M. C., Lai K. C. K. and Chen G. H. (2001). Salinity effect on mechanical dewatering of sludge with and without chemical conditioning. *Environmental Science and Technoogy*, **35**, 4691–4696.

Lv L., Yang Y., Tian J., Li Y., Li J. and Yan S. (2018). Effect of salinity on the precipitation of dissolved metals in the wastewater that produced during fly ash disposal. *IOP Conf. Series: Earth and Environmental Science*, **121**, 032012.

Marrucci G. (1969). A theory of coalescence. *Chemical Engineering Science*, **24**, 975–985.

Marrucci G. and Nicodemo L. (1967). Coalescence of gas bubbles in aqueous solutions of inorganic electrolytes. *Chemical Engineering Science*, **22**, 1257–1265.

Mesquita D. P., Amaral A. L., Ferreira E. C. and Coelho M. A. Z. (2009). Study of saline wastewater influence on activated sludge flocs through automated image analysis. *Journal of Chemical Technology and Biotechnology*, **84**, 554–560.

Mikkelsen L. H. and Keiding K. (2002). Physico-chemical characteristics of full scale sewage sludges with implications to dewatering. *Water Research*, **36**, 2451–2462.

Moon B. H., Seo G. T., Lee T. S., Kim S. S. and Yoon C. H. (2003). Effects of salt concentration on floc characteristics and pollutants removal efficiencies in treatment of seafood wastewater by SBR. *Water Science and Technology*, **47**, 65–70.

Moreno-Andrés J., Romero-Martínez L., Acevedo-Merino A. and Nebot E. (2017). UV-based technologies for marine water disinfection and the application to ballast water: Does salinity interfere with disinfection processes? *Science of the Total Environment*, **581–582**, 144–152.

Ng H. Y., Ong S. L. and Ng W. J. (2005). Effects of sodium chloride on the performance of a sequencing batch reactor. *Journal of Environmental Engineering-ASCE*, **131**, 1557–1564.

Ofori P., Firth B., Franks G., Nguyen A. and Jameson G. (2005). Impact of saline water on coal flotation. Australian Coal Association, Australia, C13051.

Oguma K., Izaki K. and Katayama H. (2013). Effects of salinity on photoreactivation of *Escherichia coli* after UV disinfection. *Journal of Water and Health*, **11**, 457–464.

Ou Y., Li R. and Liang R. (2016). Experimental study on the impact of NaCl concentration on the flocculating settling of fine sediment in static water. *Procedia Engineering*, **154**, 529–535.

Pashley R. M. and Israelachvili J. N. (1984). DLVO and hydration forces between mica Surfaces in Mg^{+2}, Ca^{+2}, Sr^{+2} and Ba^{+2} chloride solutions. *Journal of Colloid and Interface Science*, **97**, 446–455.

Portela L. I., Ramos S. and Teixeira A. T. (2013). Effect of salinity on the settling velocity of fine sediments of a harbour basin. *Journal of Coastal Research*, **65**, 1188–1193.

Prakash R., Majumder S. K. and Singh A. (2018). Flotation technique: Its mechanisms and design parameters. *Chemical Engineering and Processing: Process Intensification*, **127**, 249–270.

Pugh R. J., Weissenborn P. and Paulson O. (1997). Flotation in inorganic electrolytes; the relationship between recover of hydrophobic particles, surface tension, bubble coalescence and gas solubility. *International Journal of Mineral Processing*, **51**, 125–138.

Quinn J. J., Kracht W., Gomez C. O., Gagnon C. and Finch J. A. (2007). Comparing the effect of salts and frother (MIBC) on gas dispersion and froth properties. *Minerals Engineering*, **20**, 1296–1302.

Rasmussen H., Bruus J. H., Keiding K. and Nielsen P. H. (1994). Observations on dewaterability and physical, chemical and microbiological changes in anaerobically stored activated sludge from a nutrient removal plant. *Water Research*, **28,** 417–425.

Raynaud M., Vaxelaire J., Olivier J., Dieude-Fauvel E. and Baudez J. C. (2012). Compression dewatering of municipal activated sludge: effects of salt and pH. *Water Research*, **46**, 4448–4456.

Rincon A. G. and Pulgarin C. (2004). Effect of pH, inorganic ions, organic matter and H_2O_2 on E-coli K12 photocatalytic inactivation by TiO_2 - Implications in solar water disinfection. *Applied Catalysis B: Environmental*, **51**, 283–302.

Rubio J., Souza M. L. and Smith R. W. (2002). Overview of flotation as a wastewater treatment technique. *Minerals Engineering*, **15,** 139–155.

Rubio D., Nebot E., Casanueva J. F. and Pulgarin C. (2013). Comparative effect of simulated solar light, UV, UV/H_2O_2 and photo-Fenton treatment (UV-Vis/H_2O_2/$Fe^{2+,3+}$) in the *Escherichia coli* inactivation in artificial seawater. *Water Research*, **47**, 6367–6379.

Salvadó H., Mas M., Menéndez S. and Gracia M. P. (2001). Effects of shock loads of salt on protozoan communities of activated sludge. *Acta Protozoologica*, **40**, 177–185.

Schneider O. D., Weinrich L. A., Giraldo E. and LeChevallier M. W. (2013). Impacts of salt type and concentration on coagulation of humic acid and silica. *Journal of Water Supply: Research and Technology-Aqua*, **62**, 339–349.

Shang C., Cheung L. M., Ho C. M. and Zeng M. (2009). Repression of photoreactivation and dark repair of coliform bacteria byTiO_2-modified UV-C disinfection. *Applied Catalysis B: Environmental*, **89**, 536–542.

Sheikholeslami R. and Ong H. W. K. (2003). Kinetics and thermodynamics of calcium carbonate and calcium sulfate at salinities up to 1.5 M. *Desalination*, **157**, 217–234.

Tai C. Y. and Chen F. B. (1998). Polymorphism of $CaCO_3$, precipitated in a constant-composition environment. *AIChe Journal*, **44**, 1790–1798.

Teh C. Y., Budiman P. M., Shak K. P. Y. and Wu T. Y. (2016). Recent advancement of coagulation–flocculation and its application in wastewater treatment. *Industrial & Engineering Chemistry Research*, **55**, 4363–4389.

Tetteh E. and Rathilal S. (2019). Application of Organic Coagulants in Water and Wastewater Treatment. 10.5772/intechopen.84556.

To V. H. P., Nguyen T. V., Vigneswaran S. and Ngo H. H. (2016). A review on sludge dewatering indices. *Water Science and Technology*, **74**, 1–16.

Vaxelaire J. and Cézac P. (2004). Moisture distribution in activated sludges: a review. *Water Research*, **38**, 2215–2230.

Verwey E. J. W. and Overbeek J. Th. G. (1948). Theory of the stability of lyophobic colloids, *Amsterdam: Elsevier*.

Wang B. and Peng Y. (2013). The behaviour of mineral matter in fine coal flotation using saline water. *Fuel*, **109**, 309–315.

Wang B. and Peng Y. (2014). The effect of saline water on mineral flotation - A critical review. *Minerals Engineering*, **66**, 13–24.

Wang J., Lu H., Chen G. H., Lau G. N., Tsang W. L. and van Loosdrecht M. C. M. (2009). A novel sulfate reduction, autotrophic denitrification, nitrification integrated (SANI) process for saline wastewater treatment. *Water Research*, **43**, 2363–2372.

Wang Y., Gao B. Y., Xu X. M. and Xu W. Y. (2010). The effect of total hardness and ionic strength on the coagulation performance and kinetics of aluminum salts to remove humic acid. *Chemical Engineering Journal*, **160**, 150–156.

Wang H. T., Ye Y. Y., Qi J., Li F. T. and Tang Y. L. (2013). Removal of titanium dioxide nanoparticles by coagulation: effects of coagulants, typical ions, alkalinity and natural organic matters. *Water Science and Technology*, **68**, 1137–1143.

Wen H. J., Liu C. I. and Lee D. J. J. (1997). Size and density of flocculated sludge flocs. *Journal of Environmental Science and Health. Part A: Environmental Science and Engineering and Toxicology*, **32**, 1125–1137.

Willey J. D. (2004). The effect of ionic strength on the solubility of an electrolyte. *Journal of Chemical Education*, **81**, 1644–1646.

Winkler M. K. H., Bassin J. P., Kleerebezem R., van der Lans R. G. J. M. and van Loosdrecht M. C. M. (2012). Temperature and salt effects on settling velocity in granular sludge technology. *Water Research*, **46**, 3897–3902.

Woolard R. and Irvine R. (1995). Treatment of hypersaline wastewater in the sequencing batch reactor. *Water Research*, **29**, 1159–1168.

Wright H. and Cairns W. (1998). Ultraviolet light. Regional Symposium on Water Quality: Effective disinfection. CEPIS, Lima, pp. 1–26.

Wu G., Guan Y. and Zhan X. (2008). Effect of salinity on the activity, settling and microbial community of activated sludge in sequencing batch reactors treating synthetic saline wastewater. *Water Science and Technology*, **58**, 351–358.

Xu D., Bai Y. and Ji C. (2018). Experimental study on the settling velocity of coastal mud in quiescent water: The Case of Huangmaohai Estuary, South China Sea. *Examines in Marine Biology & Oceanography*, **1**, 61–71.

Younker J. M. and Walsh M. E. (2014). Impact of salinity on coagulation and dissolved air flotation treatment for oil and gas produced water. *Water Quality Research Journal*, **49**, 135–143.

Zhan X. (1999). Characteristics of activated sludge processes treating saline wastewater. PhD Thesis, Tsinghua University, Beijing, China.

Zhang L., Zhao Q., Wang S., Mashayekhi H., Li X. and Xing B. (2014). Influence of ions on the coagulation and removal of fullerene in aqueous phase. *Science of the Total Environment*, **466–467**, 604–608.

Zhang D., Lin Q., Xue N., Zhu P., Wang Z., Ji W. Q., Dong L., Yan K., Wu J. and Pan X. (2019). The kinetics, thermodynamics and mineral crystallography of $CaCO_3$ precipitation by dissolved organic matter and salinity. *Science of the Total Environment*, **673**, 546–552.

Zita A. and Hermansson M. (1994). Effects of ionic strength on bacterial adhesion and stability of flocs in a wastewater activated sludge system. *Applied and Environmental Microbiology*, **60**, 3041–3018.

Zuddas P. and Mucci A. (1998). Kinetics of calcite precipitation from seawater: II. The influence of the ionic strength. *Geochimica et Cosmochimica Acta*, **62**, 757–766.

Chapter 3

Salinity effects on biological treatments

ABSTRACT

High salinity environments can cause the denaturation of microbial enzymes and the destruction of the cell structure. In order to withstand these effects, microorganisms are able to develop mechanisms to both maintain the osmotic pressure balance inside the cells with the bulk liquid and retain water. Organic matter and nitrogen can be removed from high salinity effluents by means of biological processes, once microorganisms are previously adapted to the salt presence, but phosphorus removal efficiency decreases notably under saline conditions, even applying long acclimation periods or using a marine inoculum. Sudden fluctuations of wastewater salt concentration are the worst scenario for biological treatment systems since they cause the loss of pollutant removal efficiency that cannot be avoided even using halophilic microorganisms.

Keywords: Acclimation, halophilic, microorganism diversity, osmotic pressure

3.1 MOTIVATION

During wastewater treatment those effluents that contain a significant fraction of biodegradable organic matter are mostly treated by means of biological processes. These processes, based on the action of microorganisms, are sensitive to different concentrations and types of salts. To understand and determine the specific effects of salts on the different bacterial populations is of great interest for the design and operation of biological systems. In this way the size of the reactor,

Author(s): Anuska Mosquera Corral, Ángeles Val del Río and José Luis Campos Gómez

doi: 10.2166/9781789060645_0045

the treated hydraulic retention time, and the solid retention time of operation can be accurately defined to guarantee the accomplishment of the required biomass activities to produce treated effluents with the required quality for discharge or reuse.

3.2 INTRODUCTION

The performance of bacterial metabolism depends, among other factors, on the environmental osmotic pressure which is closely related to the inorganic salt concentration in the bulk liquid (Lay *et al.*, 2010). At low salt conditions, osmotic pressure inside the bacteria is higher than that of the surrounding liquid and water molecules tend to be transported inside the cells to balance the osmotic pressure. This water flux causes the microorganisms to swell and even break (turgor pressure). If salt concentration is stepwise increased, the microbial activity will be stimulated until the osmotic pressure that allows the optimal microbial activity is reached (Mosquera-Corral *et al.*, 2005). Additional increases of salt concentration will lead to the decrease of the microbial activity. According to the salt concentration range within which the microorganisms can grow under optimal conditions, they can be classified as: non-halophilic ones which stand concentrations below 10 g NaCl/L (but part of the non-halophilic microorganisms can tolerate high salt concentrations); marine or slightly halophilic ones in the range of 10–30 g NaCl/L; moderately halophilic ones in the range of 30–150 g NaCl/L; extremely halophilic ones over 150 g NaCl/L (Lay *et al.*, 2010).

Salts presence can affect the activity of the organisms, especially those that are not adapted to saline environments, in different ways (Figure 3.1):

(1) a direct effect on the enzymatic activity of the microorganisms and the subsequent deterioration of their metabolism (Cortés-Lorenzo *et al.*, 2012). At very high salt concentration, the increase of water surface tension generates competition between enzymatic proteins and salt ions for hydration and these ions strip off the essential layer of water molecules from the enzymatic proteins surface, which increases hydrophobic interactions, causing the enzyme denaturation (Sinha & Khare, 2014).

(2) at high salt concentrations an osmotic pressure gradient between the external bulk liquid and intracellular medium occurs, which causes microorganism dehydration and, in the case of extreme osmotic pressure gradients, the destruction of the bacterial cell structure (plasmolysis) (Ferrer-Polonio *et al.*, 2015).

(3) at increasing salinity, the formation of precipitates, which accumulate inside the biomass, can be induced, hindering the transfer of substrates to bacteria. This phenomenon is especially important in the case of microorganisms with low growth rates such as anaerobic and anammox bacteria (Chen *et al.*, 2008; Dapena-Mora *et al.*, 2010).

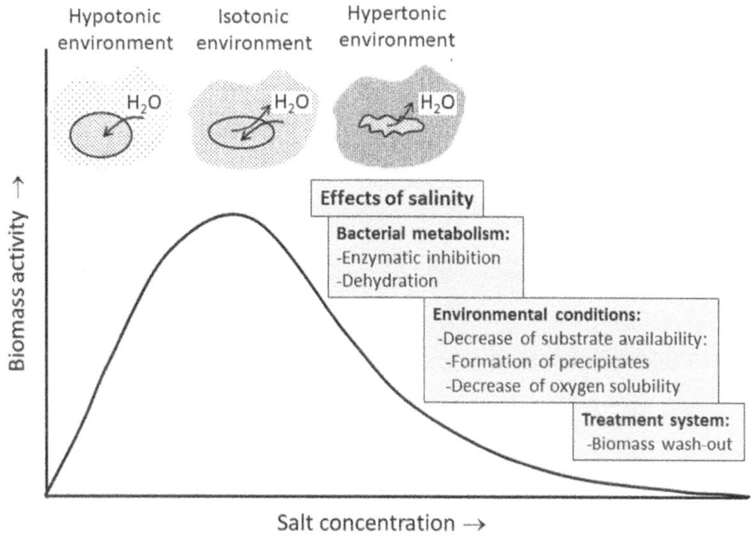

Figure 3.1 Effects of salinity on biological treatment systems.

(4) in the case of activated sludge systems the breakage of floc structure, due mainly to the inhibition of filamentous microorganism growth caused by salinity (Amin *et al.*, 2014), may lead to biomass wash-out (Salmanikhas *et al.*, 2016; Zhao *et al.*, 2016). This wash-out is also promoted by the increase of water density due to the salt presence. However, some studies suggest that this inhibition improves sludge settling properties since only dense flocs would be retained while light flocs would outflow with the effluent (Moon *et al.*, 2003; Pronk *et al.*, 2014). In any case, membrane bioreactors, biofilms or aerobic granular biomass systems can be used as alternative technologies to activated sludge systems to maintain suitable biomass concentrations under high salinity conditions (Figueroa *et al.*, 2008; Tan *et al.*, 2019).

(5) high salt concentrations reduce the solubility of oxygen and, therefore, its transfer to the liquid phase, which can lead to a decrease in biomass activity due to oxygen limitation (Bassin *et al.*, 2011).

Generally, the gradual increase of salt concentration causes an initial loss of the system removal efficiency caused by the death and/or wash-out of non-halophilic microorganisms. Then, the development of salt-tolerant microorganisms starts and, after a certain period, they become the dominant species in the sludge and the system recovers its removal efficiency (Cui *et al.*, 2015; van den Brand *et al.*, 2015) (Figure 3.2). Although this operational strategy enables a stable microbial community structure and an activated sludge with good settleability under saline

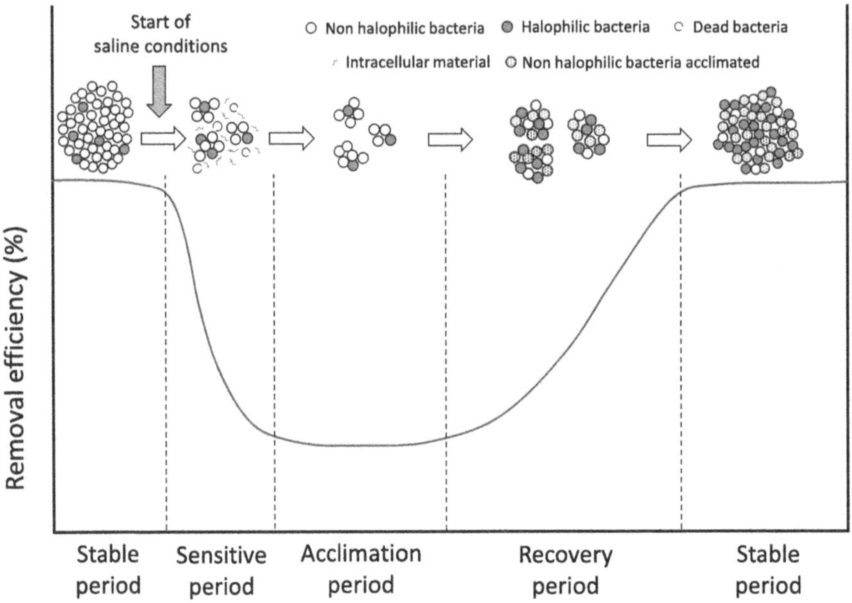

Figure 3.2 Adaptation process of a microbial community to the increase of salt concentration.

conditions to be obtained, it does not avoid the reduction of the biomass diversity as the salt concentration increases (Corsino *et al.*, 2015; Yang *et al.*, 2020). Moreover, this gradual acclimation of biomass to salt is not a useful strategy for situations, such as seawater infiltrations in coastal urban wastewater treatment plants (WWTPs) (Flood & Cahoon, 2011) and dealing with industrial wastewater (e.g. wastewater from canneries (Val del Río *et al.*, 2018) or chemical industries (Múñoz-Sierra *et al.*, 2018)), where salt concentration suffers sudden large fluctuations that make bacterial communities, especially those with a slow growth rate, unable to adapt (Lefebvre & Moletta, 2006). Even halophilic microorganisms are not able to manage these sudden fluctuations, especially sudden decreases in salt concentrations (Mcadam & Judd, 2008).

3.3 SALT TOLERANCE MECHANISMS

As mentioned above, the non-halophilic microorganisms need to adjust their metabolism as well, in order to adapt to high salt environments and in this process, they may develop new enzyme systems. However, in high salt environments, the enzyme synthesis rate is diminished, so these microorganisms need a long time to adapt to this environment (Burgess & Pletschke, 2008). Moreover, they have a limited capacity for withstanding high salinities and, over a certain salt concentration, the organic matter, nitrogen and/or phosphorus removal efficiencies of treatment systems decrease, or even are totally lost (Sharrer *et al.*, 2007; Uygur & Kargi, 2004). In this regard, bioaugmentation by using halophilic microorganisms, which already intrinsically have mechanisms to withstand high salt concentrations, has been proposed to successfully achieve the treatment of high salinity wastewaters (Kargi & Uygur, 2005).

A direct impact of the presence of high salt concentrations is the increase of osmotic pressure, causing water leakage out from bacteria. In this sense, microorganisms have developed two strategies in order to maintain an osmotic balance with their external environment (Figure 3.3) (Li *et al.*, 2018): (1) pumping inorganic ions (generally K^+) into the cell by mechanisms such as contractile vacuoles; and (2) accumulating compatible solutes from their environment or synthesizing such compatible solutes as sugars, amino acids, glycine betaine, trehalose and ectoine (Trüper & Galinski, 1990) which help to stabilize the cell membrane and cell ultrastructure (Ontiveros-Valencia *et al.*, 2014).

Since osmotic pressure is related to the contact area and surface tension, the shape of a microorganism could favour its salt tolerance. Hence, many halophilic microorganisms are rod-shaped (Tan *et al.*, 2019).

Moreover, under saline conditions, bacteria can increase their production of extracellular polymeric substances (EPS) (Li *et al.*, 2018). EPS contribute to enhancing the aggregation of the biomass but also to counteracting the osmotic pressure (Steele *et al.*, 2014). They are located surrounding the cells and, due to their negative charges (Lu *et al.*, 2019), they have the capacity for binding cations (Fang *et al.*, 2019), which are the main contributors to salt toxic actions (McCarty & McKinney, 1961). EPS also promote the water-holding capacity of bacteria (Mazor *et al.*, 1996). These facts could explain the reason why the higher the biomass aggregation, the higher the resistance under saline conditions (Guo *et al.*, 2017). Nevertheless, the increase of EPS content under saline conditions can be not only attributed to the change of microorganism metabolism but also to the microbial cell death or sludge flocs partial disintegration (Tan *et al.*, 2019).

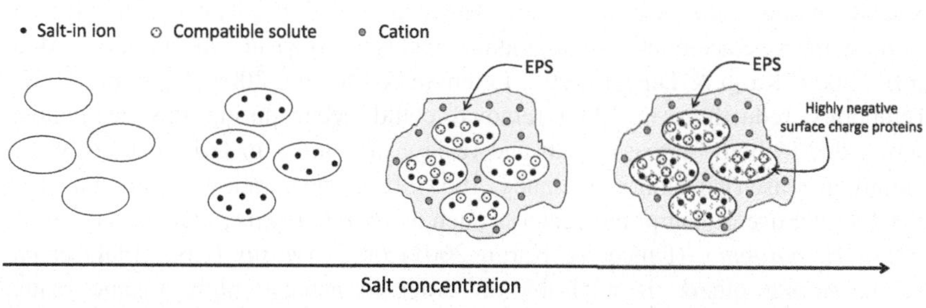

Figure 3.3 Salt tolerance mechanisms developed by microorganisms.

Both maintaining an osmotic balance and secreting EPS are high energy consumers and cause a decrease of the microorganism growth rate since the majority of the energy coming from the metabolic reactions is used for maintaining the microbial functions and, therefore, energy available to carry out cell synthesis decreases (Yogalakshmi & Joseph, 2010). This aspect is especially important in the case of treatment systems that use slow-growing microorganisms since, in these cases, the systems should be operated at high solids retentions to maintain a stable performance.

Another mechanism used by halophilic bacteria to mitigate the effects of salt stress is the synthesis of highly negative surface charge proteins which render them more flexible at high salt concentrations (Mevarech *et al.*, 2000). Other studies also showed that Gram negative species are more tolerant to salt than Gram positive ones since their outer membrane prevents some extracellular substances from penetrating the cell, which would increase its tolerance to certain toxic environments (Tan *et al.*, 2019).

3.4 ORGANIC MATTER REMOVAL

3.4.1 Aerobic treatment

The literature reviews regarding high salinity wastewater treatment show that organic matter removal under aerobic conditions can be carried out at high chemical oxygen demand (COD) removal efficiencies (>90%) (Capodici *et al.*, 2020; Lefebvre & Moletta, 2006; Tan *et al.*, 2019). Most of the reported research works tested salts concentrations up to 20–35 g NaCl/L but some studies showed that aerobic organic matter removal can successfully take place at salt concentrations as high as 80–150 g NaCl/L (Artiga *et al.*, 2008; Dincer & Kargi, 2001; Juang *et al.*, 2009; Kubo *et al.*, 2001; Lefebvre *et al.*, 2005). However, decreases of organic matter removal efficiencies under saline conditions were reported mainly due to biomass wash-out (Mirbolooki *et al.*, 2017) or to short acclimation periods (Johir *et al.*, 2013).

Previous studies also highlight that non-halophilic heterotrophic bacteria can treat low saline wastewaters without the need for an adaptation period (<10 g NaCl/L) while in the case of salinities between 10 and 40 g NaCl/L, this type of biomass must be acclimated by a gradual increase of the salt concentration (Aloui *et al.*, 2009; Kargi & Dincer, 1997; Lefebvre & Moletta, 2006; Ng *et al.*, 2005). Temporary reductions in COD removal could occur during this acclimation period due to both biomass death and release of cellular material caused by the salinity increase (Kincannon & Gaudy, 1968). For concentrations higher than 40 g NaCl/L, the use of halophilic bacteria, such as *Pseudomonas putida* (Juang *et al.*, 2009), *H. halobium* (Dincer & Kargi, 2001) or *S. xylosus* (Abou-Elela *et al.*, 2010), among others, is advisable in order to achieve high organic matter removal efficiencies.

3.4.2 Anaerobic treatment

The anaerobic digestion process takes place through four successive stages (hydrolysis, acidogenesis, acetogenesis and methanogenesis) which involve many different bacterial and archaea species, such as hydrolytic, acid forming and acetogenic bacteria, and methanogenic archaea. Studies regarding the effect of salinity on the anaerobic digestion process are mainly focused on the methanogenic stage since it is reported as the most sensitive one to the presence of salts (Mottet *et al.*, 2014; Zhao *et al.*, 2017). Generally, it is accepted that methanogenic activity is inhibited when sodium concentrations exceed 10 g/L (Chen *et al.*, 2008; De Vrieze *et al.*, 2016; Rinzema *et al.*, 1988). To overcome this issue the progressive adaptation of sludge to high salinity or the use of adapted inoculum are present as suitable strategies to achieve a stable methanogenic activity at conditions of high salt concentrations (Aspé *et al.*, 1997; Omil *et al.*, 1995). A comparison between both strategies show that similar organic loading removal rates and COD removal efficiencies can be achieved treating wastewaters with 35 g NaCl/L (Riffat & Krongthamchat, 2007). Gradual acclimation of anaerobic biomass to Na^+ enables an increase in the tolerance of methanogens to this cation. As examples, Chen *et al.* (2003) and Méndez *et al.* (1995) found that the concentration that caused an inhibition of methanogens of 100% and 90% increased from 12.7 to 22.8 g Na^+/L and from 12.0 to 17.0 g Na^+/L, respectively, after a long-term acclimation period. However, this acclimation process can take up to 700 days (Méndez *et al.*, 1995). In any case, both strategies have allowed the anaerobic process to be successfully applied for treating high salinity industrial effluents (15–71 g NaCl/L), such as seafood-processing effluents or tannery soak liquor, obtaining high efficiencies of organic matter removal (70–90%) (Guerrero *et al.*, 1997; Lefebvre *et al.*, 2006; Mosquera-Corral *et al.*, 2001; Picos-Benítez *et al.*, 2019). In order to alleviate the toxic effect of Na^+, some authors also proposed the addition of potassium ions (Li *et al.*, 2018) or exogenous compatible solutes such as glycine betaine (Zhang *et al.*, 2016) since they can be used by anaerobic microorganisms to maintain the osmotic balance with their external environment. These strategies could shorten the acclimation periods but their application at full scale would be limited by the operating costs involved.

Sulphate salts, present sometimes due to the use of seawater in the processes where wastewater is produced, can also cause problems during the anaerobic treatment of saline wastewater (Omil *et al.* 1995; Rinzema *et al.*, 1988), since sulphate reducing bacteria (SRB) compete with methanogenic microorganisms for the organic matter and produce hydrogen sulphide (H_2S). This compound inhibits the methanogenic activity by denaturing native proteins through the formation of sulphide and disulphide cross-links between polypeptide chains, interfering with the various coenzyme sulphide linkages and with the assimilatory metabolism of sulphur (Chen *et al.*, 2008). Hydrogen sulphide levels causing an activity decrease of 50% (IC_{50}) range between 50 and 250 mg H_2S/L and depend on the pH value (Chen *et al.*, 2008; Koster *et al.*, 1986). This inhibitory effect can be diminished if the biomass is previously acclimated to the saline substrate (Aspé *et al.*, 1997). Furthermore, H_2S presence in the biogas requires the implementation of specific treatment systems to remove it prior to the later use of the biogas (Okoro & Sun, 2019) while its presence in the liquid phase can be detrimental to carrying out a post-treatment to remove nitrogen by anammox based processes (Jin *et al.*, 2013) for example. For these reasons, several operating strategies were proposed to reduce the H_2S concentration inside anaerobic reactors (Guerrero *et al.*, 2016) or, even, to avoid its generation by using anaerobic purple phototrophic bacteria (Hulsen *et al.*, 2019).

During aerobic treatment, when flocculent biomass is used, the increase of the inlet salinity is correlated to the increase of the effluent turbidity and, therefore, to a worsening of effluent quality (Ali *et al.*, 2019). This could be a drawback when anaerobic membrane bioreactors are used to treat saline wastewater since membrane fouling is promoted (Muñoz-Sierra *et al.*, 2018). In this sense, granular systems can be an option to avoid operating problems during the treatment of saline wastewater since stable granules can be formed at salinities as high as 20 g Na^+/L (Gagliano *et al.*, 2017; Sudmalis *et al.*, 2018).

Regarding the effect of salinity on the anaerobic bacteria populations, most of the research works show that *Methanosaeta* is the predominant methanogen found at high salt concentrations of 59 g NaCl/L (Gagliano *et al.*, 2017; Muñoz Sierra *et al.*, 2018; Sudmalis *et al.*, 2018). On the contrary, Ali *et al.* (2019), in a reactor treating wastewater with 35 g NaCl/L, found that *Methanosarcina* was the predominant methanogen which would indicate that, in addition to salinity, there are other operational conditions that also affect the selection of the methanogenic population.

3.5 NITROGEN REMOVAL

Removal of nitrogenous compounds from wastewater is generally carried out by nitrification/denitrification processes. During nitrification, ammonia is firstly oxidized to nitrite (nitritation) by ammonia oxidizing bacteria (AOB), and subsequently the nitrite generated is converted to nitrate (nitratation) by nitrite oxidizing bacteria (NOB). Later nitrate is reduced to N_2 using organic matter by denitrifying microorganisms under anoxic conditions. Nitrification is the limiting step of conventional nitrogen removal since both AOB and NOB have low growth rates and are inhibited by the presence of both ammonia and nitrite substrates (Park & Bae, 2009). These characteristics imply that nitrifying microorganisms would be the first ones affected by both events of biomass washout and accumulation of substrates, associated with the decrease of removal efficiencies, caused by salts presence. Nevertheless, data from previous studies show that the ammonia oxidation process can perform in a stable manner up to salt concentrations of 26–56 g NaCl/L when salinity is stepwise increased (Dahl *et al.*, 1997; Panswad & Anan, 1999; Ramaswami *et al.*, 2019; She *et al.*, 2016; Vredenbregt *et al.*, 1997). Gradual adaptation of nitrifying organisms (Panswad & Anan, 1999) or proliferation of salt-tolerant strains (Chen *et al.*, 2003; Cortés-Lorenzo *et al.*, 2015) were proposed as possible mechanisms to explain the stability of ammonia oxidation at high salt concentrations.

According to literature, NOB are more sensitive to salinity than AOB but the salt concentration at which nitrite starts to be accumulated differs greatly from one research work to another (5 g NaCl/L, She *et al.* (2016); 7.6 g salt/L, Ye *et al.* (2009); 17.3 g NaCl/L, Jeong *et al.* (2018); 24.1 g NaCl/L, Cortés-Lorenzo *et al.* (2015); 24.7 g NaCl/L, Vredenbregt *et al.*, (1997)). The discrepancies among these salt concentration values could be due to the different operating conditions imposed (ammonia loading rate applied, biomass concentration, ammonia/alkalinity ratio, ammonia inlet concentration…) and/or the type of reactors used (continuous flow, Sequencing Batch Reactor (SBR), biofilm, activated sludge, …). NOB inhibition by salts presence can be used as an operational strategy to achieve short-cut nitrogen removal, by means of the nitritation process, during the treatment of both industrial and urban wastewater (She *et al.*, 2016; Song *et al.*, 2019). However, this nitrite accumulation, associated with salt presence, causes an increase of undesirable N_2O emissions (Mannina *et al.*, 2016). Moreover, NOB inhibition is not desired for example during the nitrification of aquaculture effluents when they are recycled since nitrite is much more toxic than nitrate for fish (Svobodova *et al.*, 2005). The presence of nitrite in WWTP effluents can also be a problem regarding meeting disposal requirements or when disinfection by chlorinated compounds is needed due to the increase of reagent consumption (Wahman and Speitel, 2012).

Similarly, to the nitrification process, several studies show that the denitrification process can take place under saline conditions achieving efficiencies and rates of nitrate removal comparable to those obtained at low salinity concentrations, when the salt concentration is gradually increased (Table 3.1). Moreover, Zhang *et al.* (2020) reported that the denitrification process can withstand sudden increases of salinity from 0 to 50 g NaCl/L without diminishing its efficiency of removal and observed that this salinity shock can lead to punctual nitrite accumulations which disappear after a short time of operation. This high tolerance of denitrifying bacteria to saline conditions not only depends on the bacteria type but also on the organic carbon source used as the electron donor. In this sense, Osaka *et al.* (2008) found that high nitrate removal efficiencies could be achieved at concentrations up 100 g NaCl/L when acetate was used as organic carbon, but when methanol was used, high efficiencies were maintained only up to 30 g NaCl/L.

Although, denitrifiers can withstand saline environments, the presence of high salt concentrations has a negative effect on their enzyme production system (Zhao *et al.*, 2013). In fact, the inhibition of nitrate reductase occurs

Table 3.1 Performance of denitrifying reactors treating highly and very highly saline wastewater.

Salt Concentration (g NaCl/L)	Nitrogen Loading Rate Applied (g N/L·d)	Nitrogen Removal Efficiency (%)	Wastewater Origin	Reference
16	0.06	82.7	Synthetic	Xia *et al.* (2019)
20	3.60	99	Synthetic	Jafari *et al.* (2015)
25	0.66	96.7	Synthetic	Zhu *et al.* (2015)
30	0.24	>95	Fish cannery	Capodici *et al.*, 2018
30	0.05	86.5	Pickled mustard tuber factory	Wang *et al.* (2017a)
34	0.06	96	Tannery soak liquor	Lefebvre *et al.* (2005)
34	0.30	83	Chemical factory	Ramos *et al.* (2007)
40	0.30	100	Synthetic	Zeng *et al.* (2020)
40	0.60	97	Synthetic	Zhang *et al.* (2020)
74	4.80	>95	Power plant	Vredenbregt *et al.* (1997)

above 10 g NaCl/L while the inhibition of nitrite, nitric oxide and nitrous oxide reductases starts at 40 g NaCl/L (Zeng *et al.*, 2020). Since several research works show that the nitrite reduction rate is always higher than the nitrate reduction rate under saline conditions, nitrite accumulation by the denitrification process itself is not expected (She *et al.*, 2016). However, when nitrification and denitrification processes are coupled to remove nitrogen under saline conditions, nitrite generated in the nitrifying unit enters the denitrifying unit where it could cause the inhibition of the denitrification process (Dincer & Kargi, 1999; Glass & Silverstein, 1999).

When treated wastewater does not contain enough biodegradable organic matter to carry out denitrification (for example, effluents from anaerobic digesters), nitrogen removal by means of partial nitritation/anammox processes is preferable to save operating costs related to the addition of an external source of organic matter. During partial nitritation half of the ammonia is oxidized to nitrite and then nitrite is produced and the remaining ammonia are mainly converted to N_2 under anoxic conditions by anammox bacteria. Therefore, in this case, NOB activity inhibition by salinity is desirable since this population competes for NO_2^- with anammox bacteria and causes a deterioration of the nitrogen removal efficiency (Giustinianovich *et al.*, 2018). This strategy to inhibit NOB activity could be especially useful in partial nitritation/anammox systems operated at both low nitrogen concentration and temperature (She *et al.*, 2016), since these operational conditions favour the proliferation of NOB over AOB which causes the loss of their operational stability (Gonzalez-Silva *et al.*, 2016; Pedrouso *et al.*, 2017).

On the other hand, salt stress helps to maintain the performance of both mesophilic nitrifying and anammox bacteria at temperatures up to 50°C (Courtens *et al.*, 2016; Zhang *et al.*, 2018) which opens up the possible application of autotrophic nitrogen removal processes under thermophilic conditions. This increase of thermotolerance can be attributed to: 1) the up-regulation of heat-shock proteins induced by the salt stress (Meng *et al.*, 2018); 2) salt shocks promoting the accumulation of compatible solutes inside bacteria which are osmoprotectants but also tend to inhibit the thermal denaturation of proteins by increasing their surface area (Vyrides & Stuckey, 2017); 3) salt stress enhancing the production of EPS against salt toxicity, which act against heat stress (Vyrides & Stuckey, 2009).

Several research works have focused on evaluating the effects of salinity on microbial communities operating in reactors where nitrogen compounds are removed and identified the microorganisms present under different salt concentrations (Table 3.2). The obtained results indicate that the changes of bacterial populations can be attributed not only to the salt concentration but also to the ammonia loading rate and alkalinity of the wastewater, in the case of nitrifying bacteria (Jeong *et al.*, 2018), or to the type of carbon source used, in the case of denitrification (Zhang *et al.*, 2020). In this latter case, microbial communities fed with complex organic carbon sources are more resistant to variable salinity conditions than those fed with simple substrates, such as acetate, because of their higher microbial richness and diversity (Zhang *et al.*, 2020). As a general conclusion of these studies, microbial diversity decreases with increase in salinity (Jang *et al.*, 2013; Wang *et al.*, 2019; Yoshie *et al.*, 2004) and this decrease starts at salt concentrations of 5–10 g NaCl/L (Chen *et al.*, 2003; Jang *et al.*, 2013).

3.6 PHOSPHORUS REMOVAL

Biological P-removal is mainly carried out by phosphorus accumulating organisms (PAOs) whose P content can reach up to 15% of their dry weight, while the P content stoichiometric requirement for bacterial growth only supposes 1.5–2% of their dry weight. PAOs accumulate P as polyphosphate, in aerobic or anoxic conditions,

Table 3.2 Nitrifying, denitrifying and anammox bacteria reported in literature for wastewater treatment systems operated under different salinity conditions.

Salt concentration (g NaCl/L)	AOB	NOB	Denitrifiers	Anammox
0–10	Nitrosomonas oligotropha[1] Nitrosomonas europea[1,4] Nitrosococcus mobilis[1,4] Nitrosospira sp.[1] Nitrosomonas eutropha[4] Nitrosomonas halophila[4] Nitrosomonas spp.[2,3] Nitrospira spp.[2] Nitrococcus sp.[2]	Nitrospira sp.[1] Nitrobacter spp.[3,4] Nitrospira spp.[2,3,4]	Thauera[11] Bacteroidetes[12] Azoarcus-like[12] Methylophaga[12] Paracoccus[12] Azoarcus[13]	Ca. Kuenenia stuttgartiensis[16,18,19] Ca. Scalindua wagneri[16] Ca. Jettenia sp.[17,18,23]
10–20	Nitrosomonas europea[4,10] Nitrosococcus mobilis[4,5] Nitrosospira sp.[5] Nitrosomonas eutropha[4,10] Nitrosomonas halophila[4]	Nitrobacter sp.[4,5] Nitrospira[5]	Thauera[11] Halomonas[12,14] Marinobacter[12] Azoarcus-like[12] Methylophaga[12] Paracoccus[12] Azoarcus sp.[15]	Ca. Jettenia sp.[17,18,23] Ca. Kuenenia stuttgartiensis[18,23] Unclassified Brocadiaceae[19] Ca. Scalindua sp. UK1[20]

(*Continued*)

Table 3.2 Nitrifying, denitrifying and anammox bacteria reported in literature for wastewater treatment systems operated under different salinity conditions (*Continued*).

Salt concentration (g NaCl/L)	AOB	NOB	Denitrifiers	Anammox
20–30	Nitrosospira sp.[5] Nitrosomonas europea[1] Nitrosococcus mobilis[1,4] Nitrosomonas sp.[5,6]	Nitrobacter sp.[5] Nitrospira[5] Nitrospira marina[3]	Thauera[11] Marinobacter[11,12] Halomonas[12] Azoarcus-like[12] Methylophaga[12] Paracoccus[12]	Unclassified Brocadiaceae[19] Ca. Kuenenia stuttgartiensis[16,21,22,23] Ca. Scalindua wagneri[16] Ca. Jettenia sp.[23]
30–40	Nitrosomonas sp.[6] Nitrosomonas europea[1] Nitrosomonas cryotolerands[7] N. mobilis[1,4,8] N. marina[8]	Nitrospira marina[3,7]	Marinobacter[11,12] Halomonas[12] Azoarcus-like[12] Methylophaga[12] Paracoccus[12]	
>40	Nitrosomonas europaea[8] Nitrosomonas eutropha[9] Nitrosomonas marina[8] Nitrosococcus mobilis[8]	Nitrobacter winogradskyi[9]	Halomonas[12,13] Marinobacter[12]	

[1]Moussa *et al.* (2006); [2]van Kessel *et al.* (2010); [3]Cortés-Lorenzo *et al.* (2015); [4]Chen *et al.* (2003); [5]Rejish Kumar *et al.* (2013); [6]Gonzalez-Silva *et al.* (2016); [7]Tal *et al.* (2003); [8]Cui *et al.* (2016); [9]Jeong *et al.* (2018); [10]Jang *et al.* (2013); [11]Zeng *et al.* (2020); [12]Osaka *et al.* (2008); [13]Li *et al.* (2018); [14]Yoshie *et al.* (2004); [15]Ikeda *et al.* (2013); [16]Kartal *et al.* (2006); [17]Jeong *et al.* (2020); [18]Meng *et al.* (2018); [19]Lu *et al.* (2019); [20]Kawagoshi *et al.* (2019); [21]Gonzalez-Silva *et al.* (2017); [22]Yang *et al.* (2011); [23]Jeong *et al.* (2020).

which is used as an energy reserve to uptake organic matter (as volatile fatty acids (VFAs)) under unfavourable conditions. This mechanism provides PAOs with a competitive advantage over ordinary heterotrophs. Nevertheless, glycogen accumulating organisms (GAOs) can also uptake organic matter in a similar way to PAOs, but do not accumulate P. The development of GAOs is not desirable because they compete with PAOs for the same carbon source and can decrease the phosphorus removal efficiency of WWTPs (Lopez-Vazquez *et al.*, 2009). Moreover, phosphorus can also be removed under anaerobic-anoxic alternating conditions using nitrate or nitrite as an electron acceptor, instead of oxygen, by microorganisms known as denitrifying PAOs.

Under anaerobic conditions, PAOs uptake VFAs and store them as polyhydroxyalkanoates (PHA). The energy needed for this process is obtained from polyphosphate hydrolysis or glycogen degradation. The former causes the release of phosphate into the bulk liquid. In the aerobic/anoxic phase, stored PHA is oxidized and the energy obtained is used for the P uptake and glycogen synthesis (Acevedo *et al.*, 2012). A net accumulation of P is achieved since the amount of P taken up during the aerobic/anoxic phase is higher than that released during the anaerobic phase.

In spite of having a reasonable understanding of the mechanisms associated with the biological P removal processes, there are few research works that study the effects of salinity on these processes in comparison with the amount available which evaluate such effects on organic matter and nitrogen removal processes.

The presence of salts negatively affects the activity of the PAOs and, therefore, the phosphorus removal efficiency of WWTPs. This negative effect is mainly attributed to the nitrite accumulation due to NOB activity inhibition (Bassin *et al.*, 2011; Saito *et al.*, 2004) (Figure 3.4). The accumulation of nitrite causes a decrease of both phosphate release and uptake rates during aerobic/anoxic and anaerobic phases, respectively (He *et al.*, 2019). This decrease is lower when the process takes place under anoxic conditions due to the capability of PAOs to use nitrite as an electron acceptor (Wang *et al.*, 2018). Suppression of nitrification, or even a low presence of nitrite during nitrification, reduces the negative effect of salinity, but NaCl itself still leads to a partial or total decrease in PAO activity even when long operational periods, to promote a possible biomass acclimation, or inocula from seawater are used (He *et al.*, 2019; Jiang *et al.*, 2018; Pronk *et al.*, 2014; Sharrer *et al.*, 2007; Wang *et al.*, 2017b). These research works showed discrepancies in the inhibition degree caused by NaCl, which could be related to the PAO clades present in the different experiments (Wang *et al.*, 2018), but all of them reported a decrease of PAO activity of approximately 30–40% from concentrations of 5–10 g NaCl/L. Thus, salt inhibition effects seem to be more pronounced on PAOs than on microorganisms involved in organic matter and nitrogen removal processes, probably because PAOs are more sensitive to osmotic pressure (Lefebvre & Moletta, 2006). This sensitivity is higher in the case where non-acclimated biomass is suddenly exposed to salt

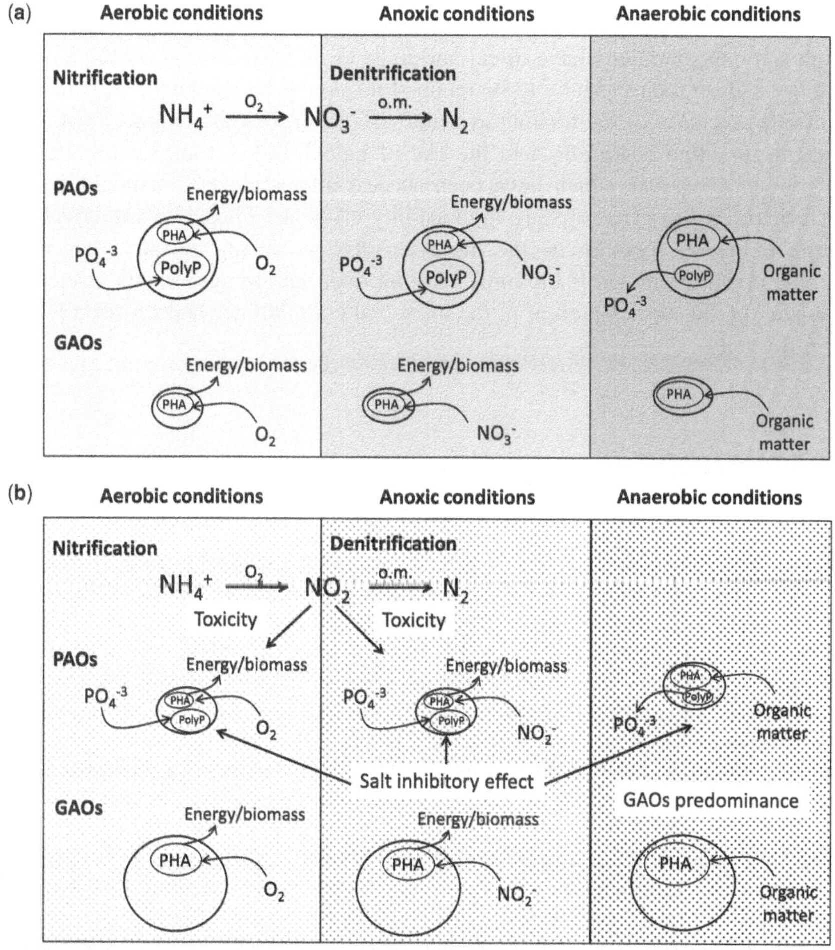

Figure 3.4 (a) Removal of P by phosphorus accumulating organisms (PAOs). (b) Effects of salinity on PAO activity (o.m.: organic matter).

since concentrations of only 1.8 g NaCl/L can cause a decrease of 63% of phosphate uptake rate (Welles *et al.*, 2015). However, it has been observed that when potassium ions are present, biological phosphorus removal can successfully occur in seawater salinity conditions, in an aerobic granular sludge system with a very high enrichment of *Ca. Accumulibacter phosphatis clade I*, since K^+ can be used by microorganisms to balance the osmotic pressure (de Graaff *et al.*, 2020).

Saline conditions also favour the development of GAOs over PAOs (He *et al.*, 2019; Welles *et al.*, 2014). Nevertheless, the shift in microbial community from PAOs to GAOs can be attributed not only to salts but also to the type of carbon source used (de Graaff *et al.*, 2020; Lopez-Vazquez *et al.*, 2009).

3.7 CONCLUSIONS

High salt concentrations have direct and indirect negative effects on the bacteria activity and on their capacity to be retained inside the bioreactors that, generally, cause the decrease of the treating systems' efficiencies. A gradual acclimation of bacteria to saline conditions and the use of halophilic microorganisms are two operational strategies which have been successfully applied to maintain a good performance of systems treating high salinity effluents. Phosphorus removal and nitrite oxidation seem to be the most sensitive processes to salt stress while aerobic organic removal is the most resistant one. Salinity also exerts a selection pressure on the species present in the microbial communities and causes a loss of their diversity.

REFERENCES

Abou-Elela S. I., Kamel M. M. and Fawzy M. E. (2010). Biological treatment of saline wastewater using a salt-tolerant microorganism. *Desalination*, **250**, 1–5.

Acevedo B., Oehmen A., Carvalho G., Seco A., Borrás L. and Barat R. (2012). Metabolic shift of polyphosphate-accumulating organisms with different levels of polyphosphate storage. *Water Research*, **46**, 1889–1900.

Ali M., Elreedy A., Ibrahim M. G., Fujii M., Nakatani K. and Tawfik A. (2019). Regulating acidogenesis and methanogenesis for the separated biogeneration of hydrogen and methane from saline-to-hypersaline industrial Wastewater. *Journal of Environmental Management*, **250**, 109546.

Aloui F., Khoufi S., Loukil S. and Sayadi S. (2009). Performances of an activated sludge process for the treatment of fish processing saline wastewater. *Desalination*, **246**, 389–396.

Amin M. M., Khiadani M. H., Fatehizadeh A. and Taheri E. (2014). Validation of linear and non-linear kinetic modeling of saline wastewater treatment by sequencing batch reactor with adapted and non-adapted consortiums. *Desalination*, **344**, 228–235.

Artiga P., García-Toriello G., Méndez R. and Garrido J. (2008). Use of a hybrid membrane bioreactor for the treatment of saline wastewater from a fish canning factory. *Desalination*, **221**, 518–525.

Aspé E., Marti M. C. and Roeckel M. (1997). Anaerobic treatment of fishery wastewater using a marine sediment inoculum. *Water Research*, **31**, 2147–2160.

Bassin J. P., Pronk M., Muyzer G., Kleerebezem R., Dezotti M. and van Loosdrecht M. C. M. (2011). Effect of elevated salt concentrations on the aerobic granular sludge process: linking microbial activity with microbial community structure. *Applied and Environmental Microbiology*, **77**, 7942–7953.

Burgess J. E. and Pletschke B. I. (2008). Hydrolytic enzymes in sewage sludge treatment: a mini-review. *Water SA*, **34**, 343–349.

Capodici M., Corsino S. F., Torregrossa M. and Viviani G. (2018). Shortcut nitrification-denitrification by means of autochthonous halophilic biomass in an SBR treating fish-canning wastewater. *Journal of Environmental Management*, **208**, 142–148.

Capodici M., Cosenza A., Di Bella G., Di Trapani D., Viviani G. and Mannina G. (2020). High salinity wastewater treatment by membrane bioreactors. In: Current Developments in Biotechnology and Bioengineering. Advanced Membrane Separation Processes for Sustainable Water and Wastewater Management – Case Studies and Sustainability Analysis, G. Mannina, A. Pandey, C. Larroche, H. Y. Ng and H. H. Ngo (eds), 1st edn, Elsevier, London, England, pp. 177–204.

Chen G. H., Wonga M. T., Okabe S. and Watanabe Y. (2003). Dynamic response of nitrifying activated sludge batch culture to increased chloride concentration. *Water Research*, **37**, 3125–3135.

Chen Y., Cheng J. J. and Creamer K. S. (2008). Inhibition of anaerobic digestion process: a review. *Bioresource Technology*, **99**, 4044–4064.

Corsino S. F., Campo R., Bella G. D., Torregrossa M. and Viviani G. (2015). Cultivation of granular sludge with hypersaline oily wastewater. *International Biodeterioration and Biodegradation*, **105**, 192–202.

Cortés-Lorenzo C., Rodríguez-Díaz M., López-López C., Sánchez-Peinado M., Rodelas B. and González-López J. (2012). Effect of salinity on enzymatic activities in a submerged fixed bed biofilm reactor for municipal sewage. *Bioresource Technology*, **121**, 312–319.

Cortés-Lorenzo C., Rodríguez-Díaz M., Sipkema D., Juárez-Jiménez B., Rodelas B., Smidt H. and González-López J. (2015). Effect of salinity on nitrification efficiency and structure of ammonia-oxidizing bacterial communities in a submerged fixed bed bioreactor. *Chemical Engineering Journal*, **266**, 233–240.

Courtens E. N., Spieck E., Vilchez-Vargas R., Bodé S., Boeckx P., Schouten S., Jauregui R., Pieper D. H., Vlaeminck S. E. and Boon N. (2016). A robust nitrifying community in a bioreactor at 50°C opens up the path for thermophilic nitrogen removal. *ISME Journal*, **10**, 2293–2303.

Cui Y. W., Su H., Chen Y. F., Chen Y. B. and Peng Y. Z. (2015). Mechanism of activated sludge floc disintegration induced by excess addition of NaCl. *Clean-Soil Air Water*, **43**, 1197–1206.

Cui Y. W., Zhang H. Y., Ding J. R. and Peng Y. Z. (2016). The effects of salinity on nitrification using halophilic nitrifiers in a Sequencing Batch Reactor treating hypersaline wastewater. *Scientific Reports*, **6**, 24825.

Dahl C., Sund C., Kristensen G. H. and Vredenbregt L. (1997). Combined biological nitrification and denitrification of high-salinity wastewater. *Water Science and Technology*, **36**, 345–352.

Dapena-Mora A., Vázquez-Padín J. R., Campos J. L., Mosquera-Corral A., Jetten M. S. M. and Méndez R. (2010). Monitoring the stability of an Anammox reactor under high salinity conditions. *Biochemical Engineeirng Journal*, **51**, 167–171.

de Graaff D. R., van Loosdrecht M. C. M. and Pronk M. (2020). Biological phosphorus removal in seawater-adapted aerobic granular sludge. *Water Research*, **172**, 115531.

de Vrieze J., Regueiro L., Props R., Vilchez-Vargas R., Jáuregui R., Pieper D. H., Lema J. M. and Carballa M. (2016). Presence does not imply activity: DNA and RNA patterns differ in response to salt perturbation in anaerobic digestion. *Biotechnology for Biofuels*, **9**, 244.

Dincer A. R. and Kargi F. (1999). Salt inhibition of nitrification and denitrification in saline wastewater. *Environmental Technology*, **20**, 1147–1153.

Dincer A. R. and Kargi F. (2001). Performance of rotating biological disc system treating saline wastewater. *Process Biochemistry*, **36**, 901–906.

Fang P. X., He X. L., Li J. F., Yang G., Wang Z. Y., Sun Z. H., Zhang X. and Zhao C. (2019). Impact of sodium ion on multivalent metal ion content in extracellular polymeric substances of granular sludge from an expanded granular sludge bed. *Environmental Technology*, **40**, 3105–3113.

Ferrer-Polonio E., Mendoza-Roca J. A., Iborra-Clar A., Alonso-Molina J. L. and Pastor-Alcañiz L. (2015). Comparison of two strategies for the start-up of a biological reactor for the treatment of hypersaline effluents from a table olive packaging industry. *Chemical Engineering Journal*, **273**, 595–602.

Figueroa M., Mosquera-Corral A., Campos J. L. and Méndez R. (2008). Treatment of saline wastewater in SBR aerobic granular reactors. *Water Science and Technology*, **58**, 479–485.

Flood J. F. and Cahoon L. B. (2011). Risks to coastal wastewater collection systems from sea-level rise and climate change. *Journal of Coastal Research*, **27**, 652–660.

Gagliano M. C., Ismail S. B., Stams A. J. M., Plugge C. M., Temmink H. and Van Lier J. B. (2017). Biofilm formation and granule properties in anaerobic digestion at high salinity. *Water Research*, **121**, 61–71.

Glass C. and Silverstein J. (1999). Denitrification of high-nitrate, high-salinity wastewater. *Water Research*, **33**, 223–229.

Giustinianovich E. A., Campos J. L., Roeckel M. D., Estrada A. J., Mosquera-Corral A. and Val del Río A. (2018). Influence of biomass acclimation on the performance of a partial nitritation-anammox reactor treating industrial saline effluents. *Chemosphere*, **194**, 131–138.

Gonzalez-Silva B. M., Jonassen K. R., Bakke I., Østgaard K. and Vadstein O. (2016). Nitrification at different salinities: biofilm community composition and physiological plasticity. *Water Research*, **95**, 48–58.

González-Silva B. M., Rønning A. J., Andreassen I. K., Bakke I., Cervantes F. J., Østgaard K. and Vadstein O. (2017). Changes in the microbial community of an anammox consortium during adaptation to marine conditions revealed by 454 pyrosequencing. *Applied Microbiology and Biotechnology*, **101**, 5149–5162.

Guerrero L., Omil F., Méndez R. and Lema J. M. (1997). Treatment of saline wastewaters from fish meal factories in an anaerobic filter under extreme ammonia concentrations. *Bioresource Technology*, **61**, 69–78.

Guerrero L., Montalvo S., Huiliñir C., Campos J. L., Barahona A. and Borja R. (2016). Advances in the biological removal of sulphides from aqueous phase in anaerobic processes: A review. *Environmental Reviews*, **24**, 84–100.

Guo Y., Liu S., Tang X. and Yang F. (2017). Role of c-di-GMP in anammox aggregation and systematic analysis of its turnover protein in Candidatus Jettenia caeni. *Water Research*, **113**, 181–190.

He Q., Wang H., Chen L., Gao S., Zhang W., Song J. and Yu J. (2019). Elevated salinity deteriorated enhanced biological phosphorus removal in an aerobic granular sludge sequencing batch reactor performing simultaneous nitrification, denitrification and phosphorus removal. *Journal of Hazardous Materials*, **390**, 121782.

Hülsen T., Hsieh K. and Batstone D. J. (2019) Saline wastewater treatment with purple phototrophic bacteria. *Water Research*, **160**, 259–267.

Ikeda N., Natori T., Okubo T., Sugo A., Aoki M., Kimura M., Yamaguchi T., Harada H., Ohashi A. and Uemura S. (2013). Enhancement of denitrification in a down-flow hanging sponge reactor by effluent recirculation. *Water Science and Technology*, **68**, 591–598.

Jafari S. J., Moussavi G. and Yaghmaeian K. (2015). High-rate biological denitrification in the cyclic rotating-bed biological reactor: Effect of COD/NO_3^-, nitrate concentration and salinity and the phylogenetic analysis of denitrifiers. *Bioresource Technology*, **197**, 482–488.

Jang D., Hwang Y., Shin H. and Lee W. (2013). Effects of salinity on the characteristics of biomass and membrane fouling in membrane bioreactors. *Bioresource Technology*, **141**, 50–56.

Jeong D., Cho K., Lee C. H., Lee S. and Bae H. (2018). Effects of salinity on nitrification efficiency and bacterial community structure in a nitrifying osmotic membrane bioreactor. *Process Biochemistry*, **73**, 132–141.

Jeong D., Kim W., Lim H. and Bae H. (2020). Shift in bacterial community structure in response to salinity in a continuous anaerobic ammonium oxidation (anammox) reactor. *International Biodeterioration and Biodegradation*, **147**, 104873.

Jiang H., Qi P., Wang T., Wang M., Chen M., Chen N., Pan L. and Chi X. (2018). Isolation and characterization of halotolerant phosphate-solubilizing microorganisms from saline soils. *3 Biotech*, **8**, 461.

Jin R. C., Yang G. F., Zhang Q. Q., Ma C., Yu J. J. and Xing B. S. (2013). The effect of sulfide inhibition on the anammox process. *Water Research*, **47**, 1459–1469.

Johir M. A. H., Vigneswaran S., Kandasamy J., Ben-Aim R. and Grasmick A. (2013). Effect of salt concentration on membrane bioreactor (MBR) performances: detailed organic characterization. *Desalination*, **322**, 13–20.

Juang R. S., Huang W. C. and Hsu Y. H. (2009). Treatment of phenol in synthetic saline wastewater by solvent extraction and two-phase membrane biodegradation. *Journal of Hazardous Materials*, **164**, 46–52.

Kargi F. and Dincer A. R. (1997). Biological treatment of saline wastewater by fed-batch operation. *Journal of Chemical Technology and Biotechnology*, **69**, 167–172.

Kargi F. and Uygur A. (2005). Improved nutrient removal from saline wastewater in an SBR by Halobacter supplemented activated sludge. *Environmental Engineering Science*, **22**, 170–176.

Kartal B., Koleva M., Arsov R., van der Star W., Jetten M. S. M. and Strous M. (2006). Adaptation of a freshwater anammox population to high salinity wastewater. *Journal of Biotechnology*, **126**, 546–553.

Kawagoshi Y., Yamashita Y., Van Duc L., Hama T. and Ito H. (2019). Changes of nitrogen-removal performance and that of the bacterial community in a mixed culture comprising freshwater and marine anammox bacteria under averaged environmental condition. *Chemosphere*, **217**, 609–617.

Kincannon D. F. and Gaudy A. F. (1968). Response of biological waste treatment systems to changes in salt concentrations. *Biotechnology and Bioengineering*, **10**, 483–496.

Koster I. W., Rinzema A., De Vegt A. L. and Lettinga G. (1986). Sulfide inhibition of the methanogenic activity of granular sludge at various pH levels. *Water Research*, **20**, 1561–1567.

Kubo M., Hiroe J., Murakami M., Fukami H. and Tachiki T. (2001). Treatment of hypersaline-containing wastewater with salt- tolerant microorganisms. *Journal of Bioscience and Bioengineering*, **91**, 222–224.

Lay W. C. L., Liu Y. and Fane A. G. (2010). Impacts of salinity on the performance of high retention membrane bioreactors for water reclamation: a review. *Water Research*, **44**, 21–40.

Lefebvre O. and Moletta R. (2006). Treatment of organic pollution in industrial saline wastewater: A literature review. *Water Research*, **40**, 3671–3682.

Lefebvre O., Vasudevan N., Torrijos M., Thanasekaran K. and Moletta R. (2005). Halophilic biological treatment of tannery soak liquor in a sequencing batch reactor. *Water Research*, **39**, 1471–1480.

Lefebvre O., Vasudevan N., Torrijos M., Thanasekaran K. and Moletta R. (2006). Anaerobic digestion of tannery soak liquor with an aerobic post-treatment. *Water Research*, **40**, 1492–1500.

Li W., Li H., Liu Y. D., Zheng P. and Shapleigh J. P. (2018). Salinity-aided selection of progressive onset denitrifiers as a means of providing nitrite for anammox. *Environmental Science and Technology*, **52**, 10665–10672.

Lopez-Vazquez C. M., Oehmen A., Hooijmans C. M., Brdjanovic D., Gijzen H. J., Yuan Z. and van Loosdrecht M. C. M. (2009). Modeling the PAO-GAO competition: effects of carbon source, pH and temperature. *Water Research*, **43**, 450–462.

Lu H., Li Y., Shan X., Abbas G., Zeng Z., Kang D., Wang Y., Zheng P. and Zhang M. (2019). A holistic analysis of anammox process in response to salinity: From adaptation to collapse. *Separation and Purification Technology*, **215**, 342–350.

Mannina G., Capodici M., Cosenza A., Di Trapani D. and Viviani G. (2016). Sequential batch membrane bio-reactor for wastewater treatment: the effect of increased salinity. *Bioresource Technology*, **209**, 205–212.

Mazor G., Kidron G. J., Vonshak A. and Abeliovich A. (1996). The role of cyanobacterial exopolysaccharides in structuring desert microbial crusts. *FEMS Microbiology Ecology*, **21**, 121–130.

Mcadam E. J. and Judd S. J. (2008). Biological treatment of ion-exchange brine regenerant for re-use: a review. *Separation and Purification Technology*, **62**, 264–272.

McCarty P. L. and McKinney R. (1961). Salt toxicity in anaerobic digestion. *Journal of the Water Pollution Control Federation*, **33**, 399–415.

Méndez R., Lema J. M. and Soto M. (1995). Treatment of seafood-processing wastewaters in mesophilic and thermophilic anaerobic filters. *Water Environment Research*, **67**, 33–45.

Meng Y., Yin C., Zhou Z. and Meng F. (2018). Increased salinity triggers significant changes in the functional proteins of ANAMMOX bacteria within a biofilm community. *Chemosphere*, **207**, 655–664.

Mevarech M., Frolow F. and Gloss L. M. (2000). Halophilic enzymes: proteins with a grain of salt. *Biophysical Chemistry*, **86**, 155–164.

Mirbolooki H., Amirnezhad R. and Pendashteh A. R. (2017). Treatment of high saline textile wastewater by activated sludge microorganisms. *Journal of Applied Research and Technology*, **15**, 167–172.

Moon B. H., Seo G. T., Lee T. S., Kim S. S. and Yoon C. H. (2003). Effects of salt concentration on floc characteristics and pollutants removal efficiencies in treatment of seafood wastewater by SBR. *Water Science and Technology*, **47**, 65–70.

Mosquera-Corral A., Sánchez M., Campos J. L., Méndez R. and Lema J. M. (2001). Simultaneous methanogenesis and denitrification of pretreated effluents from a fish canning industry. *Water Research*, **35**, 411–418.

Mosquera-Corral A., González F., Campos J. L. and Méndez R. (2005). Partial nitrification in a SHARON reactor in the presence of salts and organic carbon compounds. *Process Biochemistry*, **40**, 3109–3118.

Mottet A., Habouzit F. and Steyer J. P. (2014). Anaerobic digestion of marine microalgae in different salinity levels. *Bioresource Technology*, **158**, 300–306.

Moussa M. S., Sumanasekera D. U., Ibrahim S. H., Lubberding H. J., Hooijmans C. M., Gijzen H. J. and van Loosdrecht M. C. M. (2006). Long term effects of salt on activity, population structure and floc characteristics in enriched bacterial cultures of nitrifiers. *Water Research*, **40**, 1377–1388.

Múñoz-Sierra J. D., Oosterkamp M. J., Wang W., Spanjers H. and van Lier J. B. (2018). Impact of long-term salinity exposure in anaerobic membrane bioreactors treating phenolic wastewater: Performance robustness and endured microbial community. *Water Research*, **141**, 172–184.

Ng H. Y., Ong S. L. and Ng W. J. (2005). Effects of sodium chloride on the performance of a sequencing batch reactor. *Journal of Environmental Engineering*, **131**, 1557–1564.

Okoro O. V. and Sun Z. (2019). Desulphurisation of biogas: A systematic qualitative and economic-based quantitative review of alternative strategies. *ChemEngineering*, **3**, 76.

Omil F., Mendez R. and Lema J. M. (1995). Anaerobic treatment of saline wastewaters under high sulphide and ammonia content. *Bioresource Technology*, **54**, 269–278.

Ontiveros-Valencia A., Tang Y., Krajmalnik-Brown R. and Rittmann B. E. (2014). Managing the interactions between sulfate- and perchlorate-reducing bacteria when using hydrogen-fed biofilms to treat a groundwater with a high perchlorate concentration. *Water Research*, **55**, 215–224.

Osaka T., Shirotani K., Yoshie S. and Tsuneda S. (2008). Effects of carbon source on denitrification efficiency and microbial community structure in a saline wastewater treatment process. *Water Research*, **42**, 3709–3718.

Panswad T. and Anan C. (1999). Impact of high chloride wastewater on an anaerobic/anoxic/aerobic process with and without inoculation of chloride acclimated seeds. *Water Research*, **33**, 1165–1172.

Park S. and Bae W. (2009). Modeling kinetics of ammonium oxidation and nitrite oxidation under simultaneous inhibition by free ammonia and free nitrous acid. *Process Biochemistry*, **44**, 631–640.

Pedrouso A., Val del Río Á., Morales N., Vázquez-Padín J. R., Campos J. L., Méndez R. and Mosquera-Corral A. (2017). Nitrite oxidizing bacteria suppression based on in-situ free nitrous acid production at mainstream conditions. *Separation and Purification Technology*, **186**, 55–62.

Picos-Benítez A. R., Peralta-Hernández J. M., López-Hincapié J. D. and Rodríguez-García A. (2019). Biogas production from saline wastewater of the evisceration process of the fish processing industry. *Journal of Water Process Engineering*, **32**, 100933.

Pronk M., Bassin J. P., de Kreuk M. K., Kleerebezem R. and van Loosdrecht M. C. M. (2014). Evaluating the main and side effects of high salinity on aerobic granular sludge. *Applied Microbiology and Biotechnology*, **98**, 1339–1348.

Ramaswami S., Jalal Uddin F. M., Behrendt J. and Otterpohl R. (2019). High-rate nitrification of saline wastewaters using fixed-bed reactors. *Journal of Environmental Management*, **243**, 444–452.

Ramos A. F., Gómez M. A., Hontoria E. and González-López J. (2007). Biological nitrogen and phenol removal from saline industrial wastewater by submerged fixed-film reactor. *Journal of Hazardous Materials*, **142**, 175–183.

Rejish Kumar V. J., Sukumaran V., Achuthan C., Joseph V., Philip R. and Singh Bright I. S. (2013). Molecular characterization of the nitrifying bacterial consortia employed for the activation of bioreactors used in brackish and marine aquaculture systems. *International Biodeterioration and Biodegradation*, **78**, 74–81.

Riffat R. and Krongthamchat K. (2007). Anaerobic treatment of high-saline wastewater using halophilic methanogens in laboratory-scale anaerobic filters. *Water Environment Research*, **79**, 191–198.

Rinzema A., Van Lier J. and Lettinga G. (1988). Sodium inhibition of acetoclastic methanogens in granular sludge from a UASB reactor. *Enzyme and Microbial Technology*, **10**, 101–109.

Saito T., Brdjanovic D. and van Loosdrecht M. C. M. (2004). Effect of nitrite on phosphate uptake by phosphate accumulating organisms. *Water Research*, **38**, 3760–3768.

Salmanikhas N., Tizghadam M. and Rashidi Mehrabadi A. (2016). Treatment of saline municipal wastewater using hybrid growth system. *Journal of Biological Engineering*, **10**, 9.

Sharrer M. J., Tal Y., Ferrier D., Hankins J. A. and Summerfelt S. T. (2007). Membrane biological reactor treatment of a saline backwash flow from a recirculating aquaculture system. *Aquacultural Engineering*, **36**, 159–176.

She Z., Zhao L., Zhang X., Jin C., Guo L., Yang S., Zhao Y. and Gao M. (2016). Partial nitrification and denitrification in a sequencing batch reactor treating high-salinity wastewater. *Chemical Engineering Journal*, **288**, 207–215.

Sinha R. and Khare S. (2014). Protective role of salt in catalysis and maintaining structure of halophilic proteins against denaturation. *Frontiers in microbiology*, **5**, 165.

Song K., Sawayanagi K., Numano T., Taniichi Y., Kikuchi T., Takeda T., Kanou H., Riya S., Hori T., Hosomi M. and Terada A. (2019). High-rate partial nitrification of semiconductor wastewater: Implications of online monitoring and microbial community structure. *Biochemical Engineering Journal*, **143**, 34–40.

Sudmalis D., Gagliano M. C., Pei R., Grolle K., Plugge C. M., Rijnaarts H. H. M., Zeeman G. and Temmink H. (2018). Fast anaerobic sludge granulation at elevated salinity. *Water Research*, **128**, 293–303.

Steele D. J., Franklin D. J. and Underwood G. J. C. (2014). Protection of cells from salinity stress by extracellular polymeric substances in diatom biofilms. *Biofouling*, **30**, 987–998.

Svobodová Z., Máchová J., Poleszczuk G., Hůda J., Hamáčková J. and Kroupová H. (2005). Nitrite poisoning of fish in aquaculture facilities with water-recirculating systems. *Acta Veterinaria Brno*, **74**, 129–137.

Tal Y., Watts J. E. M., Schreier S. B., Sowers K. R. and Schreier H. J. (2003). Characterization of the microbial community and nitrogen transformation processes associated with moving bed bioreactor in a closed recirculated mariculture system. *Aquaculture*, **215**, 187–202.

Tan X., Acquah I., Liu H., Li W. and Tan S. (2019). A critical review on saline wastewater treatment by membrane bioreactor (MBR) from a microbial perspective. *Chemosphere*, **220**, 1150–1162.

Trüper H. G. and Galinski E. A. (1990). Biosynthesis and fate of compatible solutes in extremely halophilic phototrophic eubacteria. *FEMS Microbiology Letters*, **75**, 247–254.

Uygur A. and Kargi F. (2004). Salt inhibition on biological nutrient removal from saline wastewater in a sequencing batch reactor. *Enzyme and Microbial Technology*, **34**, 313–318.

Val del Rio A., Pichel A., Fernandez-Gonzalez N., Pedrouso A., Fra-Vázquez A., Morales N., Méndez R., Campos J. L. and Mosquera-Corral A. (2018). Performance and microbial features of the partial nitritation-anammox process treating fish canning wastewater with variable salt concentrations. *Journal of Environmental Management*, **208**, 112–121.

van den Brand T. P. H., Roest K., Chen G. H., Brdjanovic D. and van Loosdrecht M. C. M. (2015). Long-term effect of seawater on sulfate reduction in wastewater treatment. *Environmental Engineering Science*, **32**, 622–630.

van Kessel M. A. H. J., Harhangi H. R., van de Pas-Schoonen K., van de Vossenberg J., Flik G., Jetten M. S. M., Klaren P. H.M. and Op den Camp H. J. M. (2010). Biodiversity of N-cycle bacteria in nitrogen removing moving bed biofilters for freshwater recirculating aquaculture systems. *Aquaculture*, **306**, 177–184.

Vredenbregt L. H. J., Potma A. A., Nielsen K., Kristensen G. H. and Sund C. (1997). Fluid bed biological nitrification and denitrification in high salinity wastewater. *Water Science and Technology*, **36**, 93–100.

Vyrides I. and Stuckey D. (2009). Adaptation of anaerobic biomass to saline conditions: Role of compatible solutes and extracellular polysaccharides. *Enzyme and Microbial Technology*, **44**, 46–51.

Vyrides I. and Stuckey D. C. (2017). Compatible solute addition to biological systems treating waste/wastewater to counteract osmotic and other environmental stresses: a review. *Critical Reviews in Biotechnology*, **37**, 865–879.

Wahman D. G. and Speitel G. E. (2012). Relative importance of nitrite oxidation by hypochlorous acid under chloramination conditions. *Environmental Science and Technology*, **46**, 6056–6064.

Wang J., Gong B., Huang W., Wang Y. and Zhou J. (2017a). Bacterial community structure in simultaneous nitrification, denitrification and organic matter removal process treating saline mustard tuber wastewater as revealed by 16S rRNA sequencing. *Bioresource Technology*, **228**, 31–38.

Wang Z., van Loosdrecht M. C. M. and Saikaly P. E. (2017b). Gradual adaptation to salt and dissolved oxygen: Strategies to minimize adverse effect of salinity on aerobic granular sludge. *Water Research*, **124**, 702–712.

Wang Z., Dunne A., van Loosdrecht M. C. M. and Saikaly P. E. (2018). Effect of salt on the metabolism of 'candidatus accumulibacter' clade I and II. *Frontiers in Microbiology*, **9**, 479.

Wang H., Li H. X., Fang F., Guo J. S., Chen Y. P., Yan P. and Yang J. X. (2019). Underlying mechanisms of anammox bacteria adaptation to salinity stress. *Journal of Industrial Microbiology & Biotechnology*, **46**, 573–585.

Welles L., Lopez-Vazquez C. M., Hooijmans C. M., van Loosdrecht M. C. M. and Brdjanovic D. (2014). Impact of salinity on the anaerobic metabolism of phosphate-accumulating organisms (PAO) and glycogen-accumulating organisms (GAO). *Applied Microbiology and Biotechnology*, **98**, 7609–7622.

Welles L., Lopez-Vazquez C. M., Hooijmans C. M., van Loosdrecht M. C. M. and Brdjanovic D. (2015). Impact of salinity on the aerobic metabolism of phosphate-accumulating organisms. *Applied Microbiology and Biotechnology*, **99**, 3659–3672.

Xia Z., Wang Q., She Z., Gao M., Zhao Y., Guo L. and Jin C. (2019). Nitrogen removal pathway and dynamics of microbial community with the increase of salinity in simultaneous nitrification and denitrification process. *Science of the Total Environment*, **697**, 134047.

Yang J., Zhang L., Hira D., Fukuzaki Y. and Furukawa K. (2011). Anammox treatment of high-salinity wastewater at ambient temperature. *Bioresource Technology*, **102**, 2367–2372.

Yang Y., Wang L., Xiang F., Zhao L. and Qiao Z. (2020). Activated sludge microbial community and treatment performance of wastewater treatment plants in industrial and municipal zones. *International Journal of Environmental Research and Public Health*, **17**, 436.

Ye L., Peng C. Y., Tang B., Wang S. Y., Zhao K. F. and Peng Y. Z. (2009). Determination effect of influent salinity and inhibition time on partial nitrification in a sequencing batch reactor treating saline sewage. *Desalination*, **246**, 556–566.

Yogalakshmi K. N. and Joseph K. (2010). Effect of transient sodium chloride shock loads on the performance of submerged membrane bioreactor. *Bioresource Technology*, **101**, 7054–7061.

Yoshie S., Noda N., Tsuneda S., Hirata A. and Inamori Y. (2004). Design of 16S rRNA-targeted oligonucleotide probes and microbial community analysis in the denitrification process of a saline industrial wastewater treatment system. *FEMS Microbiology Letters*, **235**, 183–189.

Zeng D., Liang K., Guo F., Wu Y. and Wu G. (2020). Denitrification performance and microbial community under salinity and MIT stresses for reverse osmosis concentrate treatment. *Separation and Purification Technology*, **242**, 116799.

Zhang L., Zhu K. and Li A. (2016). Differentiated effects of osmoprotectants on anaerobic syntrophic microbial populations at saline conditions and its engineering aspects. *Chemical Engineering Journal*, **288**, 116–125.

Zhang Z. Z., Ji Y. X., Cheng Y. F, Xu L. Z. J. and Jin R. C. (2018). Increased salinity improves the thermotolerance of mesophilic anammox consortia. *Science of the Total Environment*, **644**, 710–716.

Zhang Z., Guo Y., Guo L., Hud F., Zhao Y., Jin C., She Z., Gao M. and Wang G. (2020). Elucidating salinity adaptation and shock loading on denitrification performance: Focusing on microbial community shift and carbon source evaluation. *Bioresource Technology*, **305**, 123030.

Zhao W., Wang Y., Liu S., Pan M., Yang J. and Chen S. (2013). Denitrification activities and N2O production under salt stress with varying COD/N ratios and terminal electron acceptors. *Chemical Engineering Journal*, **215–216**, 252–260.

Zhao Y. Y., Park H. D., Park J. H., Zhang F. S., Chen C., Li X. K., Zhao D. and Zhao F. B. (2016). Effect of different salinity adaptation on the performance and microbial community in a sequencing batch reactor. *Bioresource Technology*, **216**, 808–816.

Zhao J., Liu Y., Wang D., Chen F., Li X., Zeng G. and Yang Q. (2017). Potential impact of salinity on methane production from food waste anaerobic digestion. *Waste Management*, **67**, 308–314.

Zhu S. M., Deng Y. L., Ruan Y. J., Guo X. S., Shi M. M. and Shen J. Z. (2015). Biological denitrification using poly(butylene succinate) as carbon source and biofilm carrier for recirculating aquaculture system effluent treatment. *Bioresource Technology*, **192**, 603–610.

Chapter 4

Technologies for the treatment of saline wastewater

ABSTRACT

The existing technologies applied nowadays for the treatment of saline wastewater are presented and evaluated including physical-chemical and biological processes. Among the physical-chemical processes are conventional units like dissolved air coagulation flocculation tanks, and advanced oxidation processes where salt is not removed. Also, those processes focused on salt removal are tackled including membranes, evaporation systems, and crystallization. Salinity is responsible for the increased requirements of reagent addition during coagulation-flocculation of tertiary treatments, which incur increased operational costs. In the case of biological systems, those based on suspended and adhered biomass are considered. In general, high salt concentrations inhibit the activity of the bacterial populations responsible for the main biological processes. However, biofilm systems are supposed to be more resistant to the presence of potential inhibitors such as salts. Thus, innovative technologies based on granular/biofilm biomass are currently under study. In addition, physical-chemical processes for the separation of salts from wastewater are reviewed here.

Keywords: Inhibition, salt separation, technology

4.1 MOTIVATION

The treatment methods and technologies applied to saline wastewater can be divided into two categories: (i) those that remove/recover salt (simultaneously to other compounds) and (ii) those that remove pollutants (like organic matter and

Author(s): Anuska Mosquera Corral, Ángeles Val del Río and José Luis Campos Gómez

doi: 10.2166/9781789060645_0071

nutrients) without significantly altering the salinity of the wastewater. The first group is mainly composed of physical-chemical processes that are normally applied to streams with high salinity (>10 mS/cm) with the aim to reduce the salt concentration. The second group includes physical-chemical and biological processes. Although they are not able to reduce the salt concentration (with some exceptions), they are designed with certain modifications regarding conventional processes, to overcome the efficiency reduction due to the salt presence. As the biological processes are those affected the most by salinity, if a highly saline stream is to be treated, a pre-treatment by physical-chemical processes is recommended to remove part of the salt concentration. In this way, the inhibition of the subsequent biological process is diminished. For example, to treat wastewater from the painting industry the physical-chemical processes of nanofiltration (NF) and crystallization could be combined as a pre-treatment to remove and recover up to 80% of the sulphate in the form of crystallized Glauber salt prior to the biological treatment (Tuin *et al.*, 2006). In Figure 4.1 a decision

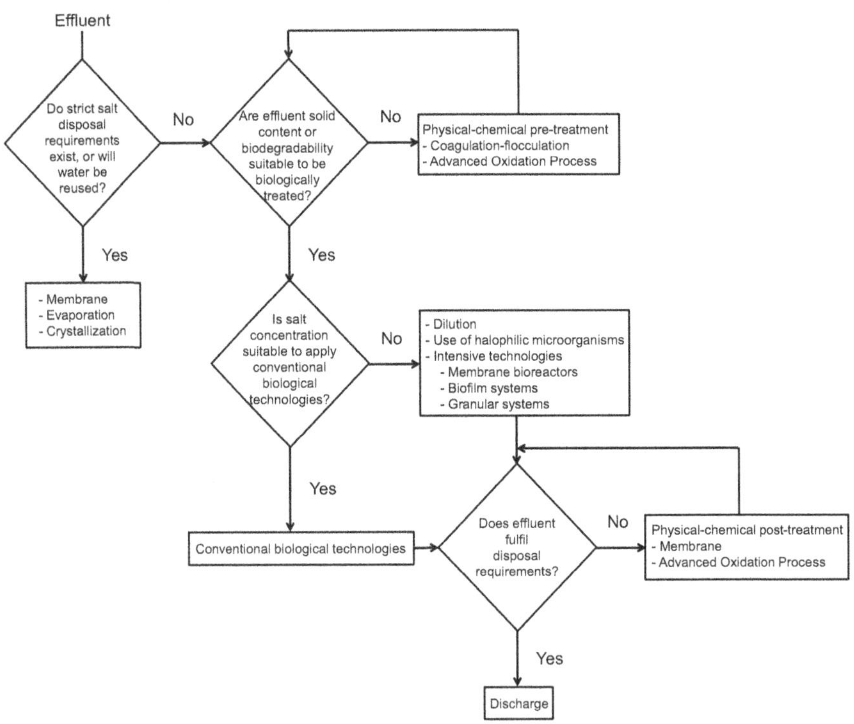

Figure 4.1 Decision tree to select the treatment technologies and processes to apply to saline streams.

tree is schematized to demonstrate the possible methods of saline wastewater treatment depending on the final disposal requirements of the treated wastewater.

Zhao *et al.* (2020), in a review of the methods for saline wastewater treatment, explain that physical-chemical technologies present some disadvantages, such as high operational costs and difficulty in achieving the expected treatment results, while biological treatment technologies have the advantage of low operational cost. Among the physical-chemical technologies, these authors highlight those based on evaporation, membrane techniques, such as reverse osmosis (RO) and nanofiltration (NF), ion exchange, advanced oxidation processes (AOPs) and electrochemical systems. In the case of biological systems, they state that the key factor for their efficient performance in treating saline wastewater is to change the environment of the bacteria to enhance their salt tolerance and halophilism. Recently, studies have mainly focused on the application of proper, cheap, and environmentally friendly biological treatment systems because of their easier maintenance and the relatively low investment and operational costs (Salmanikhas *et al.*, 2016).

In the present chapter the current technologies applied to saline wastewater treatment are presented. Although they are not applied independently, and an adequate combination of different processes is necessary to accomplish the full wastewater treatment (Figure 4.1). Examples of combined processes for complete saline wastewater treatment are described in Chapters 6 and 7, where several case studies are assessed. It is important to note that in the present chapter the units used to describe the salinity of the wastewater in the different application examples are not always the same, and variables like salt concentration, total dissolved solids (TDS) concentration and electrical conductivity (EC) are used depending on the information available in the original reference. To understand the relationship between these variables, Table 1.1 in Chapter 1 can be consulted.

4.2 PHYSICAL-CHEMICAL TECHNOLOGIES

4.2.1 Technologies for salt removal/recovery

As physical-chemical treatment processes that remove salt are very expensive, their use is limited to very specific circumstances, like some industrial streams. Additionally, it is necessary to consider that they generate a very high saline residue (brine) that sometimes contains other contaminants, like organic compounds, which should be disposed of as a hazardous waste with the corresponding cost for this disposal (Zhuo *et al.*, 2017).

In some cases, the separated salts are even recovered. Although this is not a frequent practice, it is increasing in interest nowadays due to the potential benefits of the recovered compounds (Bazedi *et al.*, 2013; Kim, 2011). More information about this aspect is available in Chapter 5.

Brine disposal cost varies from 5% to 33% of the total cost of the processes and depends on the characteristics and volume of the brine and the level of pre-treatment.

Methods of saline residue disposal include surface water discharge, sewer discharge, deep-well injection, evaporation ponds and land application (Panagopoulos *et al.*, 2019). However, the environmental impact of brine disposal and the more stringent legislation, has made it necessary to adopt certain technologies for treatment. These technologies are based on the idea of zero liquid discharge (ZLD) and comprise mainly three stages: pre-concentration, evaporation, and crystallization.

The technologies for salt removal can be classified in two main categories: membranes and thermal processes. Although specific statistics for saline wastewater treatment are not available, between both categories the data for water desalination indicate that membranes, mainly RO, comprise about 80% of the total desalination plants today worldwide, and their use continues to increase as membranes are constantly improving, both in terms of cost and energy efficiency (Anis *et al.*, 2019). Nowadays forward osmosis (FO) membranes to separate salt from wastewater are increasing in interest as they require much lower energy than RO ones for operation (Korenak *et al.*, 2017).

4.2.1.1 Membranes

Membranes have several different uses in the field of wastewater treatment, as they can be applied in both secondary biological and tertiary physical treatment systems. In general, their application is in two fields: water desalination and water reuse. Furthermore, they are the key technology in the sector of water desalination. For this reason, there is increasing research and interest in the development of effective membranes to cope with fouling problems, reduce energy consumption, increase useful life, improve maintenance, etc.

One of the major constraints of membrane processes is durability of the membranes, particularly during the treatment of wastewaters containing various surface-active contaminants (Gryta, 2020). For this reason, the use of membranes to treat saline streams is recommended when they contain low amounts of other contaminants apart from salts, to preserve the useful life of the membranes. Therefore, membrane technologies are normally applied as a tertiary treatment.

Reverse osmosis membrane systems are the most widely used to desalt saline waters and obtain a stream with low salt content, although nowadays FO is also being considered for wastewater treatment.

4.2.1.1.1 Reverse osmosis

In the RO process pressure is applied that forces the water to flow through a semi-permeable membrane from the more concentrated side (more contaminants) to the less concentrated side (fewer contaminants), resulting in the permeate having a low salt concentration and the brine containing the majority of the salts. RO systems are used not only in desalination but also in water reclamation from wastewater originating in different industrial sectors such as mining, thermal power plants, cement mills, fertilizer production, the petrochemical industry, nuclear plants, distilleries, tanneries, and production of sugar, dairy products, dyes and dye intermediates, food, pulp and paper, wood and wool, batteries, pharmaceuticals, pesticides, iron and steel, etc. (Trishitman *et al.*, 2020). The performance of RO treatments for a specific industry depends on the effluent characteristics and the pre-treatment process. Therefore, evaluation of the characteristics of the targeted effluent is needed before implementing a specific RO system.

One common example of RO application to saline industrial wastewater is in the tannery sector. Ranganathan and Kabadgi (2011) studied five tanneries operating RO membrane technologies in India. These authors observed that treatments such as neutralization, clari-flocculation and biological processes are necessary to remove organics before applying RO for TDS removal. The average inlet TDS concentration in the RO units was approximately 3,769–8,060 mg/L (average 5,530 mg/L), while in the permeate it was in the range of 118–438 mg/L (average 259 mg/L), which are values well within the drinking water standard. They observed that the removal efficiencies of TDS, sodium and chloride were in the range of 91–99%, while 70–85% of the wastewater volume was recovered and recycled in the industrial operations. Regarding the rejected brines, they were submitted to evaporation using a solar evaporation system or multiple effect evaporation technology, and the obtained solid wastes were disposed of into a secured landfill site.

Another example of RO application to wastewater reclamation is in the dairy industry. Brião *et al.* (2019) proposed the use of RO to produce water for on-site use as cooling water (permeate), while the reject stream can be reused in the same industry by adding it to fermented milk beverages and caramelized milk. Regarding the salinity removal, they observed that the salt concentration decreased from 9,480 to 321 mg/L and the conductivity from 2,200 to 95 μS/cm.

4.2.1.1.2 Forward osmosis

With respect to FO, related research studies have increased in number during the last decade. In comparison with RO the FO membranes do not require the application of hydrostatic pressure with consequent energy savings. The water is transported through a semi-permeable membrane due to differences in osmotic pressure, from the wastewater (feed solution) to a high salt concentrated draw solution. Then an additional step is necessary to separate the draw solution from the recovered water, which can be membrane distillation, a membrane bioreactor, low pressure reverse osmosis, nanofiltration, electrodialysis or thermal distillation (Valladares Linares *et al.*, 2014).

Regarding the application of FO for saline wastewater treatment there are two approaches: direct FO desalination and indirect FO desalination (Valladares Linares *et al.*, 2014). In direct FO desalination a reagent is used as the draw solution (normally $NaCl$), while in the indirect FO desalination approach the draw solution is seawater or a brackish water stream (e.g., a residual stream) with subsequent cost savings in chemicals. Several examples can be found in the literature for the direct and indirect approaches with the following combinations of wastewater-draw solution (Valladares Linares *et al.*, 2014): raw sewage plus NaCl solution; anaerobically digested sludge centrate plus $MgCl_2$ solution; secondary wastewater effluent plus NaCl solution; domestic sewage plus NaCl/$MgSO_4$ solution; drilling wastewater plus NaCl solution; wastewater from oil plus NH_4HCO_3 solution; sewage plus seawater.

The main limitation for scaling-up this technology has been identified as membrane fouling (Korenak *et al.*, 2017). Thus, different membrane materials and coatings are under exploration to decrease operational costs of maintenance. Furthermore, other critical aspects are the development of a high flux membrane, capable of maintaining an elevated salt rejection and reduced internal concentration polarization effects, and the availability of appropriate draw solutions (cost effective and non-toxic), which can be recirculated via an efficient recovery process (Valladares Linares *et al.*, 2014). A range of draw solutions has been evaluated, particularly to guarantee their regeneration capability to reduce operational costs. Although NaCl appears to be the best draw solute in terms of operation (Korenak *et al.*, 2017), KCl seems more appropriate if the aim is to use the water draw for fertigation (Sahebi *et al.*, 2020).

4.2.1.2 Evaporation

Evaporation is being considered as a good solution in an increasing number of saline wastewater treatment applications, for example it is widely used in the tannery industry. Evaporation can be effective for concentrating or removing salts, but also heavy metals and a variety of hazardous materials. Furthermore, with this technology sometimes it is possible to produce a high quality and reusable stream, which is very important when water conservation is a priority, or even to obtain the salt concentrated as a value-added sub-product. The evaporators normally used operate under vacuum conditions, so the boiling temperature of the liquid effluent is lower; thus, saving energy and improving efficiency. Vacuum evaporators operate between 45–200 mbar and 37–60°C. Modern technologies include multiple-effect evaporators, in which water is boiled in a sequence of vessels, each held at a pressure lower than that of the previous one.

On the other hand, solar evaporation ponds are a low-cost alternative, but their application is limited by land availability, potential odour problems, and meteorological and climatological conditions. These traditional evaporation ponds have a low thermal efficiency (approximately 40%), but the development of more efficient technologies based on interfacial solar evaporation, rather than on bulk water, are a promising solution as they have a much higher thermal efficiency (up to approximately 90%) (Chen *et al.*, 2019). In these systems the type of solar absorber used is crucial, and for this reason research effort is focused on the development of novel solar materials and structures to increase the evaporation rate with higher efficiency. Applications of interfacial water evaporation for wastewater purification of domestic and industrial sewage include sterilization, degradation of dye molecules, removal of heavy metal ions and separation of emulsified oil/water mixtures (Zhu *et al.*, 2019).

4.2.1.3 Crystallization

Crystallization is a solid-liquid separation technique, in which the solute crystallizes from the liquid solution and becomes a pure solid crystalline phase. This makes crystallization techniques effective for the recuperation of a variety of useful and valuable substances (along with the water) from wastewater, such as the recovery of NaCl. For example, the crystallization process can transform a salt concentrate stream (brine) into a crystalline solid that occupies little space and can be easily managed. Crystallizers are usually applied as the final treatment phase to completely dry the brine residue. However, they can be used as a single treatment process in some applications.

Lu *et al.* (2017) revised the application of the crystallization process in wastewater treatment, which mainly includes technologies like evaporation crystallization, cooling crystallization, reaction crystallization, drowning-out crystallization, and membrane distillation crystallization. These authors highlight that crystallization is one of the promising techniques for wastewater treatment, especially for water desalination and salt recovery.

4.2.2 Technologies without salt removal

Other physical-chemical technologies are not able to remove salt but are adequate for example, as pre-treatment of biological ones. This is the case for coagulation-flocculation processes and AOPs (Figure 4.1). Additionally, they have the advantage of removing pollutants (solids, recalcitrant compounds, etc.) with the corresponding beneficial effect (reduction of the inlet load) in the subsequent biological unit.

Another solution that is adopted in certain cases is the application of dilution, which is a simple option to reduce the salt concentration to overcome problems in biological processes. In this case it is advisable to have another residual stream with low organic matter, nutrient and salt concentrations but produced in large volumes (low-strength wastewater) to mix and dilute the saline stream. This option is not desirable, but common in some industrial sectors like the fish-canning industry.

4.2.2.1 Coagulation-flocculation

Coagulation-flocculation technologies can be used as a pre-treatment of saline effluents to remove their colloidal chemical oxygen demand (COD) (Lefebvre &

Moletta, 2006). The coagulation-flocculation process is based on the addition of reagents that help the suspended particles in the wastewater to form flocs able to settle. The addition of coagulants provokes the breaking down of the repulsive forces between the colloids in suspension and particles can interact with each other by agglomerating themselves in flakes that increase in size gradually. Afterwards, by the addition of flocculants, the destabilization of colloidal substances occurs to favour their sedimentation. Once flocs have reached such dimensions as to allow their precipitation, the last sedimentation phase begins, in which a calm tank allows them to settle and then be removed from the water. Therefore, the last sedimentation stage is crucial to achieve good removal efficiencies. However, the presence of salinity can reduce the settleability of the flocs formed and, consequently, specific design criteria need to be considered to improve the treatment of saline wastewater, mainly in full-scale plants operated in continuous flow. For example, Bruno *et al.* (2020) propose to include a deflector element in the passage section from the flocculation to the sedimentation compartment, which favours the sludge settling and therefore the removal of pollutants from the clarified effluent. They studied this design improvement for the application of the coagulation-flocculation process to treat saline wastewater (conductivity between 43–49 mS/cm) contaminated with hydrocarbons (1,500–1,900 mg TPH/L, as total petroleum hydrocarbons). They obtained removal efficiencies of total suspended solids (TSS) of over 90% (fed wastewater containing approximately 2,450 mg TSS/L) and of hydrocarbons close to 100%. These authors observed that the best results were obtained when the wastewater fed to the system of coagulation-flocculation was previously subjected to a treatment in which the oil was removed.

Moreover, specific studies to optimize the dosage of reagents added need to be carried out with saline effluents. For example, Di Bella *et al.* (2014) treated wastewater from an oil coastal deposit containing hydrocarbons and with a conductivity of approximately 48 mS/cm to optimize the coagulation-flocculation process. These authors concluded that the adequate pre-treatment before the biological process comprised a dose of aluminium sulphate (between 50–70 mg/L) combined with an anionic flocculant (2.5–5.0 mg/L) at neutral pH (around 7).

4.2.2.2 Advanced oxidation processes

AOPs can be employed before or after biological processes. These are popular methods for removing soluble refractory organics with wide applications in industrial wastewater treatment, and a promising solution for saline effluents. When AOPs are used as a pre-treatment stage, they can convert recalcitrant compounds into more readily biodegradable intermediates. As a post-treatment they improve effluent quality before its disposal.

For example, Moraes *et al.* (2004) demonstrated that the photo-Fenton process is feasible for the treatment of wastewaters containing hydrocarbons, even in the presence of high concentrations of salts.

Zhuo *et al.* (2017) proposed the catalytic wet peroxide oxidation (CWPO) process for saline wastewater. They explain that this process can operate under relatively mild conditions (e.g., atmospheric pressure and under 100°C), and its capital investment costs are much lower than those corresponding to other AOPs. They treated wastewater from epoxy resin synthesis, which contained total organic carbon (TOC) and salt (NaCl) concentrations of approximately 2,500–2,700 mg/L, and 18.2–20.5% by weight, respectively. In a pilot trial, at the conditions of pH = 3.3 and temperature of 90°C, these authors obtained a TOC removal efficiency of 93%. With these results they recommended treatment of the high salinity wastewater from epoxy resin synthesis with CWPO before recycling the treated wastewater to a chloro-alkali process for reuse.

4.3 BIOLOGICAL TECHNOLOGIES

4.3.1 Overview of conventional treatments

In biological systems the removal of COD and nitrogen occurs by the activity of microorganisms that can suffer from osmotic pressure if the salt concentration is above their activity threshold. If this occurs, a fraction of the microbial populations dies and consequently the pollutant removal efficiency in the treatment system diminishes. Generally, to treat saline wastewater in a conventional activated sludge (CAS) system the acclimation of the activated sludge must be accomplished by gradually increasing the salinity of the wastewater, which undoubtedly lengthens the start-up time and increases operational costs. For these reasons, the search, cultivation and domestication of salt-tolerant and halophilic microorganisms growing in extreme high-salinity environments (such as salt lakes, bay salt fields, saline-alkali soils, and pickled food factories) is a research hotspot in the treatment of saline wastewater. Halophiles prefer salt and thrive on saline or hypersaline environments. In this sense, microorganisms are classified according to their salt tolerance as true halophilic organisms if they grow in salt concentrations over 3 M NaCl, whereas salt tolerant organisms do not depend on salt for growth but can tolerate appreciable salt concentrations (over 1 M NaCl) (Zhao *et al.*, 2020).

Salt concentrations of more than 1% (NaCl) cause plasmolysis of microorganisms and reduce their ability to remove organic and inorganic pollutants. Alipour *et al.* (2016) observed that CAS systems were able to efficiently treat wastewater containing up to 4,000 mg/L of TDS (EC = 7,273 µs/cm). However, when wastewater concentration was over 4,000 mg/L of TDS, the organic matter removal in the biological treatment started to decrease and was totally lost at concentrations of approximately 8,000 mg TDS/L. In this case, the sludge settleability was scarcely affected by the salt presence. The conventional

CAS systems can be directly used to treat wastewater with salinities below 10 g NaCl/L, while the aerobic systems based in biofilms seem to be able to tolerate higher salt concentrations, because of their large biomass retention capacity (Zhao *et al.*, 2020).

Additionally, the performance of anaerobic digestion systems is also affected by salinity at mesophilic or thermophilic conditions. In fact, biomass involved in the anaerobic digestion process is more sensitive to salinity content in comparison to activated sludge (Alipour *et al.*, 2016). In view of the poor performance of anaerobic bioreactors, compatible solutes have been added into such bioreactors to help the anaerobic biomass cope with saline conditions, with for example glycine betaine (synthesized by some methanogens) being an effective one (Yang *et al.*, 2013). Salinity in special cations (i.e., Na^+) has been found to affect these processes. Anaerobic granules and flocs experience salt induced disintegration, in the presence of sodium cations, which provokes biomass washout. However, the addition of calcium ions has been proven to benefit anaerobic sludge granulation and also the strength of the formed aggregates (Pevere *et al.*, 2007). The disintegration of biomass aggregates can also be counteracted by using membrane bioreactors as proposed by Yang *et al.* (2013).

To cope with saline effluents, the use of sequencing batch reactors (SBRs) is common, as they are robust and allow the accumulation of sludge with better settling properties than that from continuous systems.

4.3.2 Halophilic microorganisms

The operation of biological processes with halophilic microorganisms for saline wastewater treatment avoids and/or reduces the inhibition by salts. In general, halophilic microorganisms can be classified as slight halophiles (1–5% NaCl), moderate halophiles (5–20% NaCl) or extreme halophiles (20–30% NaCl) (Yang *et al.*, 2013). However, the selection of adequate halophilic microorganisms can be expensive and time consuming, as they not only need to tolerate salt but also need to be able to metabolize contaminants such as organic matter, nitrogen and phosphorus. In the literature there are several examples of specific microorganisms that can be used in different conditions, however, at full scale, bioaugmentation with them is not widely used. In the following sections some examples of specific halophilic microorganisms found in the literature and applied for saline wastewater treatment are presented.

4.3.2.1 Aerobic halophiles

Kargi *et al.* (2000) performed the bioaugmentation of activated sludge with *Halobacter halobium* (a salt tolerant bacterial strain) to treat saline pickling wastewater with approximately 3.5% salt content and 4,500 mg COD/L, obtaining COD removal percentages of 95%. With the same type of industrial effluent Abou-Elela *et al.* (2010) achieved similar results with other bacterial cultures. These authors isolated a salt-tolerant microorganism (*Staphylococcus xylosus*) from a vegetable pickling plant containing about 7.2% salt and then used it combined with activated sludge to treat the wastewater from the same industry. In this study these authors managed to reach a COD removal of 93% at NaCl concentration of 3%.

4.3.2.2 Anaerobic and anoxic halophiles

Kapdan and Erten (2007) operated an upflow anaerobic packed bed reactor with synthetic saline wastewater using *Halanaerobium lacusrosei* (an anaerobic salt tolerant bacteria) as the dominant microbial culture in the process. These authors evaluated the performance of the reactor at variable concentrations of salt (0–5% NaCl) and organic matter (1,900–3,400 mg COD/L). They observed that increasing the hydraulic retention time from 11 h to 30 h resulted in a substantial improvement in the COD removal from 60 to 84% at around 3,400 mg COD/L and 3% salt concentration. A salt inhibition effect on COD removal was observed at salt concentrations above 3%.

Martínez-Espinosa *et al.* (2007) studied the application of *Haloferax mediterranei*, a denitrifying halophilic archaeon, to remove nitrate and nitrite from salted water, as it grows at concentrations as high as 2 M of nitrate and 15–25% of NaCl. They observed that this haloarchaeon could be applied to restore water tables containing high nitrate and nitrite concentrations and high salt concentrations due to fertilizer overapplication and marine intrusions, respectively.

Ali *et al.* (2020) demonstrated the application of a novel marine anammox bacterium (*Candidatus* Scalindua sp. AMX11) in a membrane bioreactor (MBR) to treat a saline effluent (of approximately 1.2% salt), achieving stable performance with a nitrogen removal rate of 0.3 kg $N/(m^3 \cdot d)$ and a removal efficiency around 90%.

4.3.2.3 Sulphate reducing bacteria

The organic matter of the wastewater is used by sulphate reducing bacteria (SRB) to transform sulphate into H_2S in anaerobic conditions. The produced H_2S is partially dissolved and partially in the biogas phase, depending on its concentration and the pH of the liquid medium. Thus, this process reduces the amount of organic matter available for methane production and causes the inhibition of the methanogenic and acetogenic bacteria, apart from being corrosive, having bad odour and contributing to the COD concentration of the produced effluent (Lens *et al.*, 1998). Although most characterized SRB strains are not well adapted to extreme hypersaline conditions some of them are halophilic organisms like *Desulfovibrio halophilus*, *Desulfocella halophila*, *Desulfovibrio oxyclinae*, and *Desulfohalobium retbaense* that have been proven to grow in environments with NaCl concentrations from 18 to 27% (Ben-Dov *et al.*, 2009; Caumette *et al.*, 1991; Kjeldsen *et al.*, 2007). In this sense, Kjeldsen *et al.* (2007) have reported moderate sulphate-reducing activity of 30 nmol/($cm^3 \cdot d$) in sediments of the Great Salt Lake (Utah) at a concentration of 27% salt, indicating the difficulty of the SRB to adapt to these conditions. For this reason, in practice H_2S production is limited by increasing salt concentrations as high as 14–18% NaCl, as observed by Ben-Dov *et al.* (2009) after adding this salt to evaporation ponds with sulphate-containing wastewater coming from several chemical plants (pesticides, pharmaceutical, and aliphatic and aromatic halogens production).

4.3.2.4 Photosynthetic microorganisms

Amezaga *et al.* (2014) conceptualized the process of desalination of wastewater by sunlight and photosynthetic organisms. These authors indicate that the future design and operation of this biodesalination process is likely to build on knowledge of both algal bioreactors and wastewater treatment biological processes. They propose the use of cyanobacterial cultures, because they can generate large amounts of biomass in brackish and seawater, thereby forming a low-salt reservoir within the saline water. The key part of the conceptualized biodesalination process is to employ a low-salt biological reservoir within the cyanobacteria as an ion exchanger. Uptake of salt into these reservoirs would then be mediated by genetic and/or environmental manipulation of the cyanobacteria. As exemplified by *Synechocystis* sp. strain PCC 6803 and *Synechococcus* sp. strain PCC 7002, cyanobacteria have attributes that make them attractive for such applications, because they are fast-growing, tolerant in a wide range of salt concentrations, and amenable to genetic transformation. Furthermore, because the primary metabolism of cyanobacteria is based upon photosynthesis, nutrient requirements are minimal and active salt export during growth is powered by sunlight. Solar radiation can also be used to energize subsequent salt accumulation through expression of retinal ion pumps. In this way, biodesalination by cyanobacteria can become an opportunity for saline wastewater treatment to reduce the salt concentration.

Hülsen *et al.* (2019) also proposed the use of purple phototrophic bacteria (PPB), for organic compounds and nutrient removal from saline and marine-like wastewaters, in a continuous anaerobic infra-red photo bioreactor. They highlighted that PPB limited electron availability for sulphate reducers when fed with sea salt (including high sulphate levels), with the advantage that no substantial H_2S was produced. The reactor microbial community shifted from *Allochromatium* sp. and *Rhodobacter* sp. to *Rhodovulum* sp. in 190 days of operation. The adapted community was halophilic, with an optimum range between 20 and 70 mS/cm, and with the capacity to maintain at least 50% of its activity at hypersaline conditions up to 140 mS/cm. These authors' results indicate a rapid adaptation to high-salt conditions (both NaCl and marine), and the capacity for PPB to form a combined wastewater treatment/resource recovery process, particularly for saline industrial wastewater.

4.3.2.5 Microalgae

Although the application of microalgae to wastewater treatment began more recently than the CAS systems, some studies have been focused on their application to saline wastewater. In this sense, Shen *et al.* (2015) stated that, compared with the screening of the organisms acclimated to saline wastewater, the utilization of marine microalgae would be more practical and economical. Marine microalgae are usually efficient in removing nitrogen and phosphorus from wastewater and can avoid contamination by biological pollutants such as bacteria, viruses and zooplankton. Additionally, microalgae can achieve potentially high yields of lipid production and high economic value of harvested biomass if further produced for fertilizers and biogas obtainment, and so on. They showed the feasibility of *Chlorella vulgaris* to remove total nitrogen (TN) and total phosphorus (TP) efficiently from real saline wastewater (EC values between 2,645–8,810 μS/cm), with a lipid accumulation of 35% at 8,810 μS/cm. Another example of microalgae application is the treatment of swine wastewater by the strains *Chlorella vulgaris*, *Botryococcus braunii* and *Desmodesmus* sp. (Nagarajan *et al.*, 2019).

4.3.3 Membrane bioreactors

The use of MBRs is a good solution for saline wastewater treatment, because biomass will be efficiently retained, providing it with a good opportunity to adapt and perform the removal activities. Furthermore, halophilic microorganisms can grow in a bioreactor due to membrane-assisted biomass retention and in this way the adverse effects of fluctuations in the influent salt concentration are minimized, thus they can also survive in environments with a wide range of salt concentrations. In addition, the retention action of the membrane allows accumulation of large biomass concentrations that compensate for the decrease of the degradation kinetics due to salt presence and, therefore, maintain the treatment capacity under saline environments. Finally, the suspended solids concentration in the effluent is low improving its quality for reuse. The use of MBR systems for saline wastewater treatment was until now applied more in aerobic conditions than in anaerobic ones (Yang *et al.*, 2013).

Tan *et al.* (2019) reviewed MBR applications for saline wastewater treatments in both, aerobic and anaerobic systems. When the salinity of the influent is below 10 g NaCl/L, the conventional MBR system can be directly used. To improve the biodegradation activities and prevent the membrane fouling at high salinities (10–100 g/L), modified and hybrid MBR systems may need to be wisely designed. Additionally, to treat hypersaline wastewaters with salinities of up to 100 g NaCl/L, inoculation of halophilic bacteria has been applied in MBR research works (Tan *et al.*, 2019).

Munz *et al.* (2008) compared the performance of an MBR with a CAS process for the treatment of tannery wastewater mixed with domestic wastewater (55–45% in volume) containing average concentrations of phenol, Cl^- and S^{-2} of 125, 3,260 and 27 mg/L, respectively (data of salinity as EC or TDS are not available). These authors' results showed that the MBR achieved COD removal efficiencies 4% higher than the CAS system and operated in a more stable manner reaching complete nitrification. Di Trapani *et al.* (2014) compared the performance of an MBR with carriers (biofilm system) and without them at increasing salinity, to study the fouling of the membrane, and concluded that the pore fouling in the membrane was higher without than with carriers. Along this line Ng *et al.* (2016) also studied the use of bio-carriers in a membrane reactor for saline pharmaceutical wastewater treatment (20,881 $\pm$ 2,030 mg/L as TDS) and observed that higher removal efficiencies of COD and TN and lower membrane fouling were obtained compared with the MBR without bio-carriers. Therefore, the combination of MBR and biofilm systems provides a good solution to treat saline wastewater.

4.3.4 Biofilm systems

The biological processes based on the use of biofilms can overcome some of the problems of the suspended ones. Although the microorganisms present in the biomass are in essence the same, their organization in biofilms can provide them with an 'extra' advantage to tackle salt inhibition due to the protective action of the biofilm structure itself. It is widely known that the biofilm structure improves the retention of biomass and can help to 'enrich' the biological system in those microorganisms that better cope with saline conditions. The biofilm can be formed over a carrier material, in which case the use of a support material is necessary, or grow by self-aggregation of microorganisms in the form of granular sludge. Below, some examples of biofilms systems applied to saline wastewater treatment are presented, to show their variability in configurations and applications.

4.3.4.1 Moving bed biofilm reactor

Different types of saline wastewater were treated using moving bed biofilm reactors (MBBRs), like agricultural drainage wastewater with salinity between 2.5–20.0 g TDS/L (Ateia *et al.*, 2015), synthetic wastewater containing phenol (200–1,200 mg/L) and salt (10–70 g NaCl/L) (Nakhli *et al.*, 2014), dairy wastewater with electrical conductivity between 2–62 mS/cm (BEST, 2020), etc. Li *et al.* (2015) compared the operation of an MBBR with a CAS system for the treatment of saline wastewater (carboxymethyl cellulose wastewater with 5–7% salinity) and they found that the MBBR performed in a more stable manner and reached higher removal rates (2.67 kg $COD/(m^3 \cdot d)$ versus 1.68 kg $COD/(m^3 \cdot d)$), being able to resist the impacts of high salinity and organic loading fluctuations.

Salmanikhas *et al.* (2016) compared the application of a hybrid growth system with a CAS system for the treatment of saline wastewater. The hybrid system consisted of a fluidized bed where polyethylene media, with an active area of

$500 \, m^2/m^3$, was introduced into a suspended biomass system. The removal percentages of COD, biological oxygen demand (BOD_5), NH_4^+, and TP have been evaluated at various salinities (0.5, 1.0 and 1.5%). The results showed that the removal performances of the hybrid system in all conditions were noticeably better than those of the CAS system. For example, they observed COD removal values of 80, 71 and 48% with the hybrid system and 62, 48 and 26% with the CAS one at the three salt concentrations tested (0.5, 1.0 and 1.5%), respectively. They attributed the better results of the hybrid system, compared to those of the CAS system, to the capacity of the former to retain the biomass under these salinity levels. Moreover, these authors identified additional advantages of the hybrid system, like no biomass recirculation requirements and a higher oxygen mass transfer coefficient for the same flow of aeration.

Another option to cope with salinity in biological treatments can be the combination of MBBR with membranes, as mentioned in the previous section. For example, Artiga *et al.* (2008) studied, at pilot scale, a hybrid biofilm-suspended biomass membrane bioreactor to treat wastewater from a fish-canning factory (with salt concentration up to 84 g/L). They observed the adaptation of the biomass to salinity after 73 days and obtained COD removal efficiencies of approximately 92%.

4.3.4.2 Anaerobic granular sludge

In the frame of the project SaltGae (SALTGAE, 2020) successful results were obtained with a two-phase anaerobic digestion system, with separated acidogenic and methanogenic stages, using granular sludge. The results showed that the produced biogas was rich in methane (up to 85%) and, therefore, it could be used for energy generation. The acidogenic stage allowed treatment of saline wastewaters containing up to 58 g Na^+/L, while the methanogenic stage was more sensitive and allowed production of up to 0.23 L $CH_4/(L \cdot d)$ at a salinity up to 14 g Na^+/L. These studies report the highest salinities that an acidogenic system can cope with to date and demonstrate its potential to obtain valuable products, like volatile fatty acids, from saline wastewater.

4.3.4.3 Aerobic granular sludge

Aerobic granular sludge (AGS) for saline wastewater treatment was also used by diverse authors as an adequate technology to combat the salt inhibitory effects over biological processes. Carrera *et al.* (2019) studied the application of different configurations (fully aerobic or anaerobic feeding) for the treatment of fish-canning wastewater with salinities between 5.0–13.5 g NaCl/L. In this study not only was the increase of salinity addressed but also, as a novelty, the effect of salinity decrease was tested. The results demonstrated that when the salt concentration decreased (from 10 to 5 g NaCl/L) the stability of the granular aggregates in the anaerobic feeding reactor was worse than in the aerobic feeding one; while when the salinity increased (from 5.5 to 13.5 g NaCl/L) in both systems a loss of biomass retention occurred. Additionally, this study demonstrated the robustness of the AGS technology to treat saline wastewater with removal efficiencies around 80–90% treating organic loading rates between 2.0–6.7 kg $COD_s/(m^3 \cdot d)$.

Recently, Sarvajith and Nancharaiah (2020) cultivated AGS directly from the autochthonous planktonic bacterial community of seawater. They used an SBR directly fed with nutrient-amended seawater without inoculum. They obtained dense and compact granules with excellent settling velocity (41.2 m/h) in few days (25 d). The microbiological analysis showed an enrichment in α-proteobacteria and that the bacterial diversity of the granules was lower than the seawater microbiome. The halophilic AGS was able to efficiently remove nitrogen biologically and treat seafood-processing wastewater without previous biomass acclimation. The ammonium and total nitrogen removals were 92–100% and 81–90%, respectively, while nitrite accumulation was not observed under saline conditions. Phosphate removal progressively improved and stabilized at over 76% within 2 months from the start-up.

4.3.5 Wetlands

Liang *et al.* (2017) state that for the treatment of saline wastewater, constructed wetlands (CWs) outperform many other processes. These systems have received increasing attention in recent years, especially in developing countries, because of their minimal costs, convenient operation, eco-friendly characteristics, and aesthetic value. However, the treatment of industrial saline wastewater by CWs has been tested in limited cases which include the effluents generated in sectors such as the tannery, textile, pulp and paper industry, and aquaculture ponds (Liang *et al.*, 2017). In these cases, to achieve adequate removal rates, it is necessary to use specific plants adapted to salinity (e.g., halophytes or salt-tolerant plants). Furthermore, in CWs treatment the role of the microorganism is essential. It was demonstrated that CWs inoculated with specific halophilic microorganisms provide a promising method for treating saline wastewater. Although salts might pose some negative effects on the performance of CWs, past research suggests that the halophytes and halophilic microorganisms that exist in natural saline or hypersaline environments can be introduced into CWs to improve the treatment of saline wastewater. An acceptable removal efficiency can be achieved in CWs by using a combination of halophytes, halophilic microorganisms, and optimal substrates.

4.4 OTHER INNOVATIVE TECHNOLOGIES

The growing interest in addressing the increasing salt concentrations present in different types of wastewater, is expected to make the development of new technologies and processes feasible. Some examples of these new developments are presented in this section. Most of them are still at an early stage of maturation, without tests at industrial scale, but they are promising options which are worthy of consideration.

4.4.1 Micro-electrolysis

For high salinity wastewater, micro-electrolysis with iron-carbon alloy appears to be a feasible technology to remove salt. Zhou *et al.* (2015) proposed the pre-treatment of saline wastewater with a Fe–C alloy filler prepared using iron powder and activated carbon. These authors tested this technology to treat modified starch wastewater (23,520 mg COD/L, 20,880 mg Cl^-/L), pharmaceutical wastewater (5,890 mg COD/L, 21,500 mg Cl^-/L) and cellulose ether wastewater (12,400 mg COD/L, 15,360 mg Cl^-/L). They observed COD and Cl^- removals of over 55% and approximately 25–30%, respectively. These authors explained that although activated carbon in the filler had physical adsorption ability, Cl^- was absorbed mainly through electric-absorption caused by the primary cell reaction of the Fe–C alloy. The oxygen consumption curves of the analysed industrial samples showed that the biodegradability of wastewater could be greatly improved by the removal of Cl^- using the new filler.

4.4.2 Temperature swing solvent extraction

In 2019 a novel process was proposed to remove salt from hypersaline streams called temperature swing solvent extraction (TSSE) (Boo *et al.*, 2019). This process can treat water that contains a high concentration of salts, even up to seven times saltier than seawater. This methodology is proposed as a radically different desalination technology because it is membrane-less and not based on evaporative phase change. TSSE operates at relatively low temperature (less than 70°C) and uses a low-polarity solvent (with temperature-dependent water solubility) for the selective extraction of water over salt. The preliminary results showed that it was able to remove up to 98.4% of the salt, which is comparable with reverse osmosis. But unlike reverse osmosis or other methods of desalination, this process does not require high temperatures or high pressures.

4.4.3 Microbial desalination cell

Zhang *et al.* (2019) proposed the microbial desalination cell (MDC) as a novel and promising technology for wastewater desalination. The MDC can produce energy from wastewater and simultaneously desalinate it driven by bioelectricity generation. The authors state that compared to traditional desalination treatment processes, MCD shows certain advantages, such as less energy consumption and more mild operating conditions. They studied the biocathode MDC for simultaneous salt, sulphate, and nitrogen removal from high-salinity mustard tuber wastewater (approximately 20 g TDS/L and EC of approximately 30–35 mS/cm). The results showed that 97.4% of salt, 99.7% of sulphate and 99.8% of nitrogen could be removed. The relative abundances of electrogenic bacteria in the anode and cathode were 15.95% and 15.10%, respectively, which greatly promoted the electricity generation and desalination. The bacteria involved in sulphate reduction in the anode were the dominant population, with relative abundance of 13.94%. Microbial community analysis of the cathode biofilm indicated that autotrophic nitrification-anaerobic denitrification, electrochemical reduction, and anaerobic ammonium oxidation might coexist for high-efficiency nitrogen removal.

Yang *et al.* (2019) revised the application of MDC and other bioelectrochemical systems (BESs) to saline wastewater treatment, indicating that to guarantee an adequate treatment producing high quality effluents for potential reuse and to achieve energy self-sufficiency, these systems have been upgraded by integrating membrane-based treatment technologies. These integrated BESs have been expanded to desalination applications and have exhibited improved energy recovery and wastewater treatment efficiencies compared to conventional BESs.

4.4.4 Microbial mats

Microbial mats are clustered biofilms of mixed microbial communities that occur naturally in hypersaline habitats, with abundant microbial diversity. Akyon *et al.* (2015) used engineered microbial mats for the treatment of saline hydraulic fracturing produced water. They observed the highest biodegradation rates between 0 and 50,000 mg TDS/L in the treated water. When the concentration was 100,000 mg TDS/L this rate decreased, and no biological substrate removal was detected for values of 200,000 mg TDS/L. The authors stated that microbial mats have the potential to decrease operational costs and improve the efficiency of treating saline produced water from hydraulic fracturing. Therefore, microbial mats represent an emerging biological treatment technology to encourage produced water reuse, improve the performance of physical-chemical treatment approaches, remove organic constituents, and reduce chemicals application.

4.5 EXAMPLES OF TREATMENT SCHEMES

The characteristics of the saline wastewater and the disposal or reuse requirements strongly affect the type of treatment that needs to be applied (Figure 4.1). In Table 4.1, some examples of treatment trains are presented selected from different types of industrial saline wastewater (landfill, tannery, textile, refinery, and fish flour/oil industry) together with the different technologies available. It is important to highlight that biological treatments are applied in all these examples, emphasizing their relevance for COD and nutrients removal, even in the case of the refinery wastewater. Another important aspect is that energy intensive processes like ultrafiltration and reverse osmosis are only applied as a post-treatment when the treated water is going to be reused.

One interesting example is the treatment of landfill leachates from the Green Lane Landfill in St. Thomas (Ontario, Canada). This stream is characterized by

Table 4.1 Examples of technologies applied for the treatment of different types of saline wastewater.

Type of Effluent	Country	Characteristics	Pre-treatment	Biological Process	Post-treatment	Reference
Landfill leachate	Canada	COD (1,000–5,000 mg/L), ammonia (150–600 mg/L), TDS (2,000–6,000 mg/L) and intense colour (up to 1,200 Pt-Co)	Coagulation-flocculation and primary clarification for metals, hardness, and TDS removal	Modified Ludzack-Ettinger (MLE) process	Ozonation to remove colour and filtration to remove solids	Arabi and Lugowski (2015)
Tannery	Ethiopia	pH (8), conductivity (21 mS/cm), COD (2,100 mg/L), sulfide (191 mg S/L), sulfate (571 mg S/L), ammonia (32 mg N/L)	Aerated equalization pond for sulfide oxidation	Aerated biological oxidation pond for COD removal	Coagulation (for organics, phosphate, and chromium removal)	Abate *et al.* (2020)

(*Continued*)

Table 4.1 Examples of technologies applied for the treatment of different types of saline wastewater (*Continued*).

Type of Effluent	Country	Characteristics	Pre-treatment	Biological Process	Post-treatment	Reference
Textile	India	pH (8–9), COD (970 mg/L), TDS (3,500 mg/L), colour (3,536 Hazen)	Equalization tank and pH adjustment	Aerobic process with specific microbial culture added each month	Coagulation-flocculation Activated carbon filter (removal of colour/odour) Ozonation Ultrafiltration Reverse osmosis (50% permeate is reused in the industry)	Nakhate *et al.* (2020)
Refinery	India	pH (8–9), oil and grease (110–220 mg/L), COD (930 mg/L), 280 (mg/L)	Primary oil separation by floatation Secondary oil separation by coagulation-flocculation	Bio-tower packed with polypropylene for removal of toxic hydrocarbons. Aeration tank for biodegradable organic matter removal	Pressure sand filters and granular activated carbon filters for removal of turbidity and colour.	Ashfaq *et al.* (2010)
Fish flour and fish oil industry	Ecuador	600–800 mg N/L, 30–80 mg P/L, 5,000–8,000 mg COD/L	Cooling tower and equalization tank	Aerobic membrane bioreactor with nitrification-denitrification	Ultrafiltration and reverse osmosis, for water reuse and irrigation	Fluence (2020)

high solid and colloidal organics concentrations, as heavy metals, and xenobiotic organic compounds. The treatment train with a capacity of 300 m^3/d includes a pre-treatment with chemicals addition and primary clarification to remove metals, hardness, and TDS. Subsequently, there is a biological process and as final step consecutive ozonation (to remove colour) and filtration (to remove solids) steps. The complete treatment train achieves removal percentages of 60–80%, 60–70% and 70–90% for COD, TN, and colour, respectively.

4.6 CONCLUSIONS

Technologies applied for the treatment of saline wastewater are mostly the same as those applied for non-saline ones. In many cases only changes in the operational conditions are applied to overcome the drawbacks exerted by the salinity. Nevertheless, in the case of the physical-chemical treatments specific technologies are applied for the separation of the salt fraction, such as the conventional membrane, evaporation, and crystallization technologies or innovative technologies like micro-electrolysis, temperature swing solvent extraction, microbial desalination cells or microbial mats. Additionally, when selecting the biological technology to be used, some options are considered to improve the biomass retention and for this purpose membrane systems or the development of biofilm or granular sludge are implemented. Also, inoculation with halophilic cultures has been explored, as a beneficial option to speed up the start-up and maintain the biological process stability, or the use of systems like wetlands.

However, due to the increase in the volume of wastewater with saline conditions produced in the world, it would be an interesting practice to evaluate the development of specific technologies designed from their concept to deal with saline wastewater.

REFERENCES

Abate T. A., Feleke Desta A. and Love N. G. (2020). Evaluating tannery wastewater treatment performance based on physicochemical and microbiological characteristics: an Ethiopian case study. *Water Environment Research*, **1364**, 1–12.

Abou-Elela S. I., Kamel M. M. and Fawzy M. E. (2010). Biological treatment of saline wastewater using a salt-tolerant microorganism. *Desalination*, **250**, 1–5.

Akyon B., Stachler E., Wei N. and Bibby K. (2015). Microbial mats as a biological treatment approach for saline wastewaters: the case of produced water from hydraulic fracturing. *Environmental Science and Technology*, **49**(10), 6172–6180.

Ali M., Shaw D. R. and Saikaly P. E. (2020). Application of an enrichment culture of the marine anammox bacterium 'Ca. Scalindua sp. AMX11' for nitrogen removal under moderate salinity and in the presence of organic carbon. *Water Research*, **170**, 115345.

Alipour V., Moein F. and Rezaei L. (2016). Determining the salt tolerance threshold for biological treatment of salty wastewater. *Health Scope*, **6**(1), e36425.

Amezaga J. M., Amtmann A., Biggs C. A., Bond T., Gandy C. J., Honsbein A., Karunakaran E., Lawton L., Madsen M. A., Minas K. and Templeton M. R. (2014). Biodesalination: A case study for applications of photosynthetic bacteria in water treatment. *Plant Physiology*, **164**, 1661–1676.

Anis S. F., Hashaikeh R. and Hilal N. (2019). Reverse osmosis pretreatment technologies and future trends: A comprehensive review. *Desalination*, **452**, 159–195.

Arabi S. and Lugowski A. (2015). Lessons learned from successful applications of biological landfill leachate treatment. Environmental & Science Engineering Magazine, https://esemag.com/biosolids/lessons-learned-successful-applications-biological-landfill-leachate-treatment/ (accessed 6 November 2020).

Artiga P., García-Toriello G., Méndez R. and Garrido J. M. (2008). Use of a hybrid membrane bioreactor for the treatment of saline wastewater from a fish canning factory. *Desalination*, **221**, 518–525.

Ashfaq A., Saadia A., Anees S. and Khursheed A. (2010). Environmental waste management: A case study of refinery. *Journal of Insdustrial Pollution Control*, **26**(1), 95–103.

Ateia M., Nasr M., Yoshimura C. and Fujii M. (2015). Organic matter removal from saline agricultural drainage wastewater using a moving bed biofilm reactor. *Water Science and Technology*, **72**(8), 1327–1333.

Bazedi G. A., Ettouney R. S., Tewfik S. R., Sorour M. H. and El-Rifai M. A. (2013). Salt recovery from brine generated by large-scale seawater desalination plants. *Desalination and Water Treatment*, **52**, 4689–4697.

Ben-Dov E., Kushmaro A. and Brenner A. (2009). Long-term surveillance of sulfate-reducing bacteria in highly saline industrial wastewater evaporation ponds. *Saline Systems*, **5**, 1–5.

BEST (2020). Effectiveness of moving bed biofilm reactor (MBBR) technology for industrial wastewater treatment and impact of sodium chloride on the biomass in MBBR. Summary of Project BEST output (WP2.2). Riga Technical University. https://bestbalticproject.eu/wp-content/uploads/2020/05/MBBR-piloting_BEST_RTU.pdf (accessed 6 November 2020).

Boo C., Winton R. K., Conway K. M. and Yip N. Y. (2019). Membrane-less and non-evaporative desalination of hypersaline brines by temperature swing solvent extraction. *Environmental Science and Technology Letters*, **6**, 359–364.

Brião V. B., Vieira Salla A. C., Miorando T., Hemkemeier M. and Cadore Favaretto D. P. (2019). Water recovery from dairy rinse water by reverse osmosis: giving value to water and milk solids. *Resources Conservation and Recycling*, **140**, 313–323.

Bruno P., Campo R., Giustra M. G., De Marchis M. and Di Bella G. (2020). Bench scale continuous coagulation-flocculation of saline industrial wastewater contaminated by hydrocarbons. *Journal of Water Process Engineering*, **34**, 101156.

Carrera P., Campo R., Méndez R., Di Bella G., Campos J. L., Mosquera-Corral A. and Val del Rio A. (2019). Does the feeding strategy enhance the aerobic granular sludge stability treating saline effluents? *Chemosphere*, **226**, 865–873.

Caumette P., Cohen Y. and Matheron R. (1991). Isolation and characterization of *Desulfovibrio halophilus* sp. No., a halophilic sulfate-reducing bacterium isolated from Solar Lake (Sinai). *Systematic and Applied Microbiology*, **14**, 33–38.

Chen C., Kuang Y. and Hu L. (2019). Challenges and opportunities for solar evaporation. *Joule*, **3**(3), 683–718.

Di Bella G., Giustra M. G. and Freni G. (2014). Optimisation of coagulation/flocculation for pre-treatment of high strength and saline wastewater: performance analysis with different coagulant doses. *Chemical Engineering Journal*, **254**, 283–292.

Di Trapani D., Di Bella G., Mannina G., Torregrossa M. and Viviani G. (2014). Comparison between moving bed-membrane bioreactor (MB-MBR) and membrane bioreactor (MBR) systems: influence of wastewater salinity variation. *Bioresource Technology*, **162**, 60–69.

Fluence (2020). Aerobic Treatment of Wastewater for Fish Flour and Fish Oil Company for Reuse. https://www.fluencecorp.com/wp-content/uploads/2018/12/REUSE_Case-study_Tadel_Eng_lores.pdf (accessed 06 November 2020).

Gryta M. (2020) Separation of saline oily wastewater by membrane distillation. *Chemical Papers*, **74**, 2277–2286.

Hülsen T., Hsieh K. and Batstone D. J. (2019). Saline wastewater treatment with purple phototrophic bacteria. *Water Research*, **160**, 259–267.

Kapdan I. K. and Erten B. (2007). Anaerobic treatment of saline wastewater by *Halanaerobium lacusrosei*. *Process Biochemistry*, **42**(3), 449–453.

Kargi F., Dinner A. R. and Pala A. (2000). Characterization and biological treatment of pickling industry wastewater. *Bioprocess Engineering*, **23**, 371–374.

Kim D. H. (2011). A review of desalting process technique and economic analysis of the recovery of salts from retentates. *Desalination*, **270**, 1–8.

Kjeldsen K. U., Loy A., Jakobsen T. F., Tomsen T. R., Wagner M. and Ingvorsen K. (2007). Diversity of sulfate-reducing bacteria from an extreme hypersaline sediment, Great Salt Lake (Utah). *FEMS Microbiology Ecology*, **60**, 287–298.

Korenak J., Basu S., Balakrishnan M., Hélix-Nielsen C. and Petrinic I. (2017). Forward osmosis in wastewater treatment processes. *Acta Chimica Slovenica*, **64**, 83–94.

Lefebvre O. and Moletta R. (2006). Treatment of organic pollution in industrial saline wastewater: A literature review. *Water Research*, **40**(20), 3671–3682.

Lens P. N. L., Visser A., Janssen A. J. H., Hulshoff Pol L. W. and Lettinga G. (1998). Biotechnological treatment of sulfate-rich wastewaters. *Critical Review in Environmental Science and Technology*, **28**(1), 41–88.

Li K., Xu J., Yun G., Zhao Y., Dong Z., Liu J. and Wang K. (2015). Treatment of high salt wastewater in MBBR process strengthened by complex salt-tolerant microorganism. *Chinese Journal of Environmental Engineering*, **9**(6), 2829–2834.

Liang Y., Zhu H., Bañuelos G., Yan B., Zhou Q., Yu X. and Cheng X. (2017). Constructed wetlands for saline wastewater treatment: a review. *Ecological Engineering*, **98**, 275–285.

Lu H., Wang J., Wang T., Wang N., Bao Y. and Hao H. (2017). Crystallization techniques in wastewater treatment: an overview of applications. *Chemosphere*, **173**, 474–484.

Martínez-Espinosa R. M., Zafrilla B., Camacho M. and Bonete M. J. (2007). Nitrate and nitrite removal from salted water by *Haloferax mediterranei*. *Biocatalysis and Biotransformation*, **25**(2–4), 295–300.

Moraes J. E. F., Quina F. H., Nascimento C. A. O., Silva D. N. and Chiavone-Filho O. (2004). Treatment of saline wastewater contaminated with hydrocarbons by the photo-Fenton process. *Environmental Science and Technology*, **38**(4), 1183–1187.

Munz G., Gualtiero M., Salvadori L., Claudia B. and Lubello C. (2008). Process efficiency and microbial monitoring in MBR (membrane bioreactor) and CASP (conventional activated sludge process) treatment of tannery wastewater. *Bioresource Technology*, **99** (18), 8559–8564.

Nagarajan D., Kusmayadi A., Yen H. W., Di D. C., Lee D. J. and Chang J. S. (2019). Current advances in biological swine wastewater treatment using microalgae-based processes. *Bioresource Technology*, **289**, 121717.

Nakhate P. H., Moradiya K. K., Patil H. G., Marathe K. V. and Yadav G. D. (2020). Case study on sustainability of textile wastewater treatment plant based on lifecycle assessment approach. *Journal of Cleaner Production*, **245**, 118929.

Nakhli S. A. A., Ahmadizadeh K., Fereshtehnejad M., Rostami M. H., Safari M. and Borghei S. M. (2014). Biological removal of phenol from saline wastewater using a moving bed biofilm reactor containing acclimated mixed consortia. *Springerplus*, **3**(112), 24616843.

Ng K. K., Shi X., Ong S. L., Lin C. F. and Ng H. Y. (2016). An innovative of aerobic bio-entrapped salt marsh sediment membrane reactor for the treatment of high-saline pharmaceutical wastewater. *Chemical Enineering Journal*, **295**, 317–325.

Panagopoulos A., Haralambous K.-J. and Loizidou M. (2019). Desalination brine disposal methods and treatment technologies – a review. *Science of the Total Environment*, **693**, 133545.

Pevere A., Guibaud G., van Hullebusch E. D., Boughzala W. and Lens P. N. L. (2007). Effect of Na^+ and Ca^{2+} on the aggregation properties of sieved anaerobic granular sludge. *Colloids and Surfaces A*, **306**, 142–149.

Ranganathan K. and Kabadgi S. D. (2011). Studies on feasibility of reverse osmosis (membrane) technology for treatment of tannery wastewater. *Journal of Environmental Protection*, **2**, 37–46.

Sahebi S., Sheikhi M., Ramavandi B., Ahmadi M., Zhao S., Adeleye A. S., Shabani Z. and Mohammadi T. (2020). Sustainable management of saline oily wastewater via forward osmosis using aquaporin membrane. *Process Safety and Environmental Protection*, **138**, 199–207.

Salmanikhas N., Tizghadam M. and Rashidi Mehrabadi A. (2016). Treatment of saline municipal wastewater using hybrid growth system. *Journal of Biological Engineering*, **10**, 9.

SaltGae (2020). Project Results SaltGae Stakeholders Platform. http://www.stakeholder saltgae.eu/knowledge-transfer/project-results/ (accessed 6 November 2020).

Sarvajith M. and Nancharaiah Y. V. (2020). Granulation of the autochthonous planktonic bacterial community of seawater for saline wastewater treatment. *Environmental Science: Water Research and Technology*, **6**, 1902–1916.

Shen Q. H., Gong Y. P., Fang W. Z., Bi Z. C., Cheng L. H., Xu X. H. and Chen H. L. (2015). Saline wastewater treatment by *Chlorella vulgaris* with simultaneous algal lipid accumulation triggered by nitrate deficiency. *Bioresource Technology*, **193**, 68–75.

Tan X., Acquah I., Liu H., Li W. and Tan S. (2019). A critical review on saline wastewater treatment by membrane bioreactor (MBR) from a microbial perspective. *Chemosphere*, **220**, 1150–1162.

Trishitman D., Cassano A., Basile A. and Rastogi N. K. (2020). 9 – Reverse osmosis for industrial wastewater treatment. In: Current Trends and Future Developments of (Bio-) Membranes, A. Basile, A. Cassano and N. K. Rastogi (eds), Elsevier, pp. 207–228.

Tuin B. J. W., Geerts R., Westerink J. and van Ginkel C. G. (2006). Pretreatment and biotreatment of saline industrial wastewaters. *Water Science and Technology*, **53**(3), 17–25.

Valladares Linares R., Li Z., Sarp S., Bucs S. S., Amy G. and Vrouwenvelder J. S. (2014). Forward osmosis niches in seawater desalination and wastewater reuse. *Water Research*, **66**, 122–139.

Yang J., Spanjers H., Jeison D. and Van Lier J. B. (2013). Impact of Na^+ on biological wastewater treatment and the potential of anaerobic membrane bioreactors: A review. *Critical Reviews in Environmental Science and Technology*, **43**(24), 2722–2746.

Yang E., Chae K. J., Choi M. J., He Z. and Kim I. S. (2019). Critical review of bioelectrochemical systems integrated with membrane-based technologies for desalination, energy self-sufficiency, and high-efficiency water and wastewater treatment. *Desalination*, **452**, 40–67.

Zhang L., Fu G. and Zhang Z. (2019). High-efficiency salt, sulfate and nitrogen removal and microbial community in biocathode microbial desalination cell for mustard tuber wastewater treatment. *Bioresource Technology*, **289**, 121630.

Zhao Y., Zhuang X., Ahmad S., Sung S. and Ni S. Q. (2020). Biotreatment of high-salinity wastewater: current methods and future directions. *World Journal of Microbiology and Biotechnology*, **36**, 37.

Zhou G., Wang X., Liu J., Wang Z. and Li S. (2015). Pretreatment of saline wastewater with Fe–C alloy filler. *Desalination and Water Treatment*, **53**(3), 603–609.

Zhu L., Gao M., Peh C. K. N. and Ho G. W. (2019). Recent progress in solar-driven interfacial water evaporation: advanced designs and applications. *Nano Energy*, **57**, 507–518.

Zhuo Y., Sheng M., Liang X. and Cao G. (2017). Treatment of high salinity wastewater using CWPO process for reuse. *Journal of Advanced Oxidation Technologies*, **20**(2), 20170024.

Chapter 5

Valorization of saline wastewater

ABSTRACT

In the current frame of the circular economy, evaluation of the valorization potential of saline-type wastewater is an objective, as this type of effluent represents approximately 9% of all the produced wastewater in the world. Apart from the evident water recovery for reuse, accomplishing valorization of the organic and inorganic pollutant compounds is also explored. Obtaining products, such as volatile fatty acids (VFA), exopolymers (EPS) or polyhydroxyalkanoates (PHA) among others, from organic carbon compounds, entails the development of innovative processes, which must produce treated effluents suitable for discharge or reuse. In addition, the organic compounds can be anaerobically converted to methane-containing biogas to be used as an energy source. Concerning the inorganic fraction, either nutrients or salts recovery offers an opportunity to extract some economic value from these effluents. Furthermore, there are industrial processes admitting relatively high saline effluents in their units, such as some refrigeration towers, that allow for direct water reuse.

Keywords: Biopolymer, energy, nutrient, resource recovery, water reclamation

5.1 MOTIVATION

In the frame of the circular economy concept, wastewater perception is moving from considering it as a waste to seeing it as a resource of valuable compounds (Figure 5.1). With this general perspective the use of resources will be more

Author(s): Anuska Mosquera Corral, Ángeles Val del Río and José Luis Campos Gómez

doi: 10.2166/9781789060645_0097

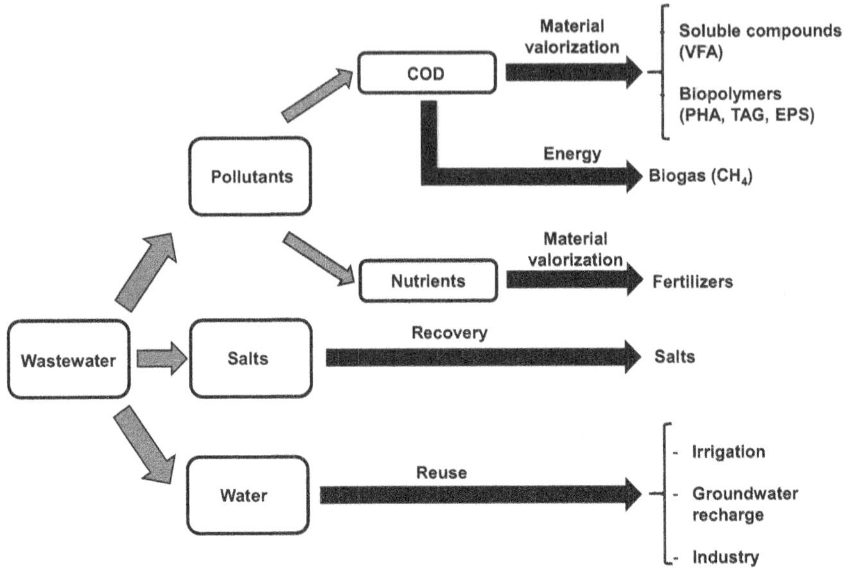

Figure 5.1 Valorization alternatives for wastewater. COD: chemical oxygen demand, VFA: volatile fatty acids, PHA: polyhydroxyalkanoates, TAG: triacylglycerides, EPS: exopolysaccharides.

efficient and the production of waste minimized while some materials are recovered (Puyol *et al.*, 2016). Thus, wastewater management becomes more sustainable and environmentally friendly as the concept shift from what must be removed to what can be recovered is accomplished (Guest *et al.*, 2009).

Wastewater, from urban or industrial origin, contains compounds worthy of recovering, such as cellulose fibres (Ruiken *et al.*, 2013), organic matter to produce biopolymers (Lin *et al.*, 2010) like bioplastics (Reis *et al.*, 2003) or building blocks such as volatile fatty acids (VFA) (Kleerebezem *et al.*, 2015), nutrients such as phosphorus (Crutchik *et al.*, 2017), energy as methane to move towards wastewater treatment self-sufficiency (Rajeshwari *et al.*, 2000) or heavy metals (Cervantes and Pavlostathis, 2006). Most of the wastewater streams presented in Chapter 1 are potential sources for resource recovery and, for this reason, some alternatives for their valorization are discussed here.

Apart from the recovery of valuable compounds, the reuse of the water itself is increasing in importance nowadays (Guest *et al.*, 2009). Population growth together with the associated requirements for food production is creating a critical situation with respect to water availability worldwide. Moreover, groundwater quality is diminishing in some cases, due to freshwater over abstraction in coastal areas and the consequent seawater intrusion, reaching saline concentrations as high as 5–10 g/L, or associated with the leaching and migration of fertilizers from irrigated agricultural zones (Horriche and Benabdallah, 2020). Besides, climate change is reducing the flows of freshwater which are the basis for the replenishment of those water basins where human water resources are stored. In certain places the water availability is so limited that freshwater supply requires its transportation from long distances and even from different countries (Guest *et al.*, 2009). Recognition should be given to the importance of access to water for economic activity development that can be covered by reused wastewater (Jodar-Abellan *et al.*, 2019).

For these reasons, new strategies need to be applied to mitigate or reduce the water consumption in such a way that its access is guaranteed to all the inhabitants in the world. Different alternatives to diminish freshwater consumption have been put into practice beginning with optimizing the use of freshwater in households and industry. Also, dual water supply systems comprising freshwater for potable uses and seawater to flush toilets have been applied in places like Hong Kong where the latter amounts to 20–30% of the water required in the entire city (Liu *et al.*, 2016). Among these strategies, the reuse of regenerated/reclaimed wastewater is arising in importance. However, saline wastewater needs specific evaluation to establish its potential for reuse as its salt content might constitute a drawback for certain uses.

5.2 BY-PRODUCTS OBTAINMENT

Productive sectors such as the chemical, petroleum, pharmaceutical, leather manufacture, agro-food, and aquaculture industries, among others, produce wastewater characterized by large salt contents. These effluents are saline or hypersaline streams that amount to 5% of the produced industrial effluents (Lefebvre and Moletta, 2006). Additionally, the fact that in some places, such as Hong Kong, seawater is used for urban purposes, like toilet flushing, increases the production of this type of effluent (Liu *et al.*, 2016).

Frequently, saline wastewater also contains large concentrations of organic matter and nutrients. These pollutants may also be a source of valuable compounds. If the organic fraction is transformed into organic polymeric compounds or biogas, useful as substitutes for those from chemical origin, the current pressure on petroleum-based materials might be alleviated. This new concept is feasible as these obtained biopolymers are a source of biofuel, commodities for the biochemical industry and biomaterials (Venkata Mohan *et al.*, 2016), and biogas contains methane which can be used for energy production. In addition, simultaneously to the waste valorization, the accurate treatment of these streams needs to be performed to guarantee achieving the appropriate characteristics of the effluent for discharge or reuse.

5.2.1 Organic compounds production

Valorization of the organic matter contained in wastewater streams can be achieved by transforming it into valuable products like volatile fatty acids and biopolymers, produced intra- and extracellularly (Mosquera-Corral *et al.*, 2017). Most of these processes are in a development stage and their application is not widespread. Albeit they are expected to be implemented in the future. A common issue for all the compounds to be recovered from wastewater is that the recovery/extraction step needs to be appropriately selected to obtain products with the required quality for application (Atasoy *et al.*, 2019). But this issue is outside of the scope of this book.

5.2.1.1 Volatile fatty acids (VFA)

The production of short chain VFA is increasing in interest due to their potential applications. These compounds can be used as building blocks for the chemical industry (Kleerebezem *et al.*, 2015) or methane generation, as a carbon source for further biological processes like nutrient removal or biopolymer production, or for biodiesel production or electricity generation in microbial fuel cells (Atasoy *et al.*, 2019; Zhou *et al.*, 2018).

Anaerobic digestion (AD) is a biological process applied generally to wastewater containing large biodegradable organic matter concentrations. VFA are intermediate compounds of anaerobic digestion, which has methane as the final product. In the AD process VFA are formed in the acidogenic step before the methanogenesis. Thus, VFA accumulation is an undesired event in those anaerobic systems intended for biogas production, which suffer from hydraulic or organic overloads. Additionally, in other cases VFA accumulation is a result of the presence of inhibitory compounds like large salt concentrations (de Vrieze *et al.*, 2017). Thus, the production of VFA relies on the definition of the accurate operational conditions that stop the complete development of the AD process in the acidogenic or acetogenic steps. The spectrum of the mixture of VFA obtained depends on operational conditions such as pH, temperature, flow rate, type and concentration of substrate, and VFA concentration or feeding pattern, among others (Yin *et al.*, 2016). Studies are under development now to elucidate the relationship between these parameters and the obtained VFA mixture (Atasoy *et al.*, 2018).

Wastewater of industrial origin containing large organic matter concentrations is a potential source for VFA production (Table 5.1). The use of saline media for this purpose has not been widely explored yet (Fra-Vázquez *et al.*, 2020; Scoma *et al.*, 2017). In general, when dealing with waste streams, mixed microbial consortia are preferred. This is due to their associated low operational cost compared to pure cultures as a result of the non-sterile conditions required, variable biochemical functions, applicability to different substrates, large operational flexibility, and easy process control and scale up (Dahiya *et al.*, 2015). These microbial consortia produce different mixtures of VFA containing mainly acetic (HAc), propionic (HPr), butyric (HBu) and valeric acids (HVa), among others. Among the operational conditions, the optimal pH reported in the literature varies widely and

Table 5.1 VFA production from saline industrial wastewater streams.

Waste Origin	OLR g COD/(L · d)	Temperature (°C)	pH	HRT (d)	Yield g COD_{VFA}/g $COD_{NVFArem}$	Max COD to VFA (%)	Salt (g/L)	Reference
Molasses[1]	9.81	34	7	10	86.8	63.5	20 000–90 000*	Scoma *et al.* (2017)
Mussel cooking process	2.5 ± 0.4	37 ± 1	4.2	6.3		43	21.8 ± 2.8 g NaCl/L	Fra-Vázquez *et al.* (2020)

*In electrical conductivity units (μS/cm).

[1]Average values of experiments performed with terrestrial and marine inoculum.

OLR: organic loading rate; $COD_{NVFArem}$: non-volatile fatty acids as COD removed.

depends mostly on the type of wastewater. This parameter ranges from approximately 5.25–5.50 for cheese whey (Bengtsson *et al.*, 2008), to 7.0 for kitchen wastes (Zhang *et al.*, 2005), and as high as 10–11 for activated sludge (Chen *et al.*, 2007).

When saline effluents are treated, special attention needs to be paid to biomass development together with an appropriate adjustment of the operational conditions and salt concentrations for VFA accumulation. VFA were produced from molasses, by microbial communities without acclimation, up to conductivities of 90 000 μS/cm in continuously operated packed-bed biofilm reactors (PBBRs) filled with ceramic carriers (Scoma *et al.*, 2017).

Previous studies indicate that the presence of salts does not negatively affect the VFA accumulation, achieving conversion rates like those obtained with low saline effluents (Scoma *et al.*, 2017). Even low pH values allowed for the obtainment of 43% conversion of soluble COD to VFA in a reactor fed with mussel cooking wastewater with concentrations of 22 g NaCl/L (Fra-Vázquez *et al.*, 2020). In this study the major VFA was acetic acid followed by butyric in the initial operational stages and propionic at the end of the experimental period. The limited obtained conversion was due to the presence of proteins which were not converted due to the low pH of operation.

5.2.1.2 Biopolymers

Under certain limiting operational conditions, microorganisms produce certain compounds as extra- or intra-cellular biopolymers (Mosquera-Corral *et al.*, 2017). Extracellular compounds, named exopolymeric substances (EPS), have mainly structural or protective functions for the biomass, in many cases helping the formation of biofilms. Additionally, cells accumulate intracellular compounds to store carbon and energy that are used in conditions of the absence of available substrate (Thomson *et al.*, 2010). These compounds include polyhydroxyalkanoates (PHA), triacylglycerides (TAG) and polyglucose (PG) among others.

5.2.1.2.1 Exopolymeric substances (EPS)

Microorganisms excrete exopolymeric substances in conditions of stress such as the presence of salts, etc. (Chen *et al.*, 2018b). Thus, EPS have a preserving function on the microbial cells protecting them from the presence of damaging environmental conditions like inhibitory compounds. These adverse environments include the existence of high salinity conditions from which these EPS protect the cells (Chen *et al.*, 2018b; Zeng *et al.*, 2016). Although this effect does not always occur. In the case of anaerobic granular biomass, the salinity hinders the production of EPS (Ismail *et al.*, 2010).

Most of the EPS are produced independently of the salinity of the wastewater. Those associated with the presence of salt are not frequent. In this sense, Xue *et al.* (2019) obtained bioactive sulphated polysaccharides (SP) (fucoidan, carrageenan and heparin) from activated sludge collected from laboratory and full-scale reactors treating saline sewage containing sulphates. SP are sulphur-containing macromolecules widely used as industrial feedstock. The yields obtained by Xue *et al.* (2019) were of 342.8 $\pm$ 0.3 and 418.1 $\pm$ 0.4 g SP/kg VSS (VSS stands for volatile suspended solids) from sludge coming from a laboratory and a full-scale reactor, respectively. The types and contents of the obtained SP were highly influenced by the operating conditions that need to be further assessed to develop the recovery process.

Although information about these processes applied to saline wastewater is not available for all products obtainment the potential for EPS to be generated from saline wastewater has been already proven (Ajao *et al.*, 2019).

5.2.1.2.2 Polyhydroxyalkanoates (PHA)

Polyhydroxyalkanoates are compounds with physical-chemical properties like those of the plastics from petroleum origin. They are generally co-polymers of hydroxybutyrate (HB) and hydroxyvalerate (HV) monomers at different ratios. PHA are biodegradable and they are produced from bacterial cultures and using organic compounds, like certain wastes, as a substrate. For this reason, they are called bioplastics and are expected to substitute conventional plastics in the future.

Many types of microorganisms accumulate PHA in the interior of the cells as storage compounds, which serve as internal energy and carbon reserves, to be used when organic matter is not available in the liquid media. PHA are produced in aerobic conditions mainly from VFA, in preference to butyrate, lactate or acetate. In this way VFA that are in liquid phase are accumulated inside the microbial cells. This is a procedure which concentrates the organic matter present in the wastewater. However, using substrates like glycerol means the previous VFA production can be omitted as this compound can be used directly for PHA storage in a mixed microbial culture (Moralejo-Gárate *et al.*, 2011).

The production of PHA in halophilic conditions has been studied extensively in systems based on pure cultures (Quillaguamán *et al.*, 2010). Nevertheless, information is scarce regarding PHA obtainment from saline effluents and with non-pure cultures (Argiz *et al.*, 2020; Palmeiro-Sánchez *et al.*, 2016a, 2016b) (Table 5.2). The use of halophilic mixed cultures presents several advantages. On the one hand, non-sterile conditions are feasible, as saline conditions hinder the activity of competing non-halophilic organisms (Tan *et al.*, 2011), and waste streams can be used as a substrate. On the other hand, the fact that PHA-accumulating mixed cultures grow in halophilic conditions is beneficial as it may facilitate further biopolymer extraction by provoking the cells lysis after exposing them to distilled water, which avoids the use of solvents (Pramanik *et al.*, 2012). In these conditions operational costs decrease compared to those processes based on pure cultures. However, some drawbacks need to be overcome when dealing, for example, with saline fish canning wastewater, such

Table 5.2 PHA accumulation studies with microbial mixed cultures (MMC) using saline industrial wastewater streams previously acidified. Provided results correspond to the accumulation step.

Waste Fed	NaCl (g/L)	Temperature (°C)	Yield		HB: HV (%)	Max. PHA (wt%)	Reference
			Cmol HB/Cmol VFA	Cmol HV/Cmol VFA			
Synthetic media	0	30	0.529	0.191	2.71	53	Palmeiro-Sánchez *et al.* (2016a)
	7		0.462	0.164	3.13	34.6	
	13		0.523	0.128	4.67	17.4	
	20		0.556	0.063	6.37	8.9	
Pre-acidified cooked mussel processing industry	12	30	0.538	0.192	70:30	60	Argiz *et al.* (2020)

HB: Hydroxybutyrate; HV: Hydroxyvalerate; PHA: Polyhydroxyalkanoates.

as the fact that the amount of stored biopolymer decreases due to the presence of salt and the monomers composition significantly changes (Palmeiro-Sánchez *et al.*, 2016a).

In all the research studies using saline waste effluents, the PHA production took place using mixed microbial cultures (MMC). Furthermore, the applied process is a three-unit system (Valentino *et al.*, 2017). This system comprises an acidification unit to produce the VFA mixture. The obtained VFA-rich stream is then the substrate for the further two steps: the second one to enrich the PHA-accumulating mixed culture and the third one to evaluate the maximal PHA-accumulation capacity of the previous mixed culture (Mosquera-Corral *et al.*, 2017).

The enrichment of the MMC can take place by applying two different strategies: i) applying aerobic dynamic feeding to establish the feast/famine regime by alternating conditions of the presence/absence of substrate in the liquid media (Johnson *et al.*, 2009); ii) alternating anaerobic/aerobic conditions to select for polyphosphate- (PAO) and glycogen-accumulating organisms (GAO) (Serafim *et al.*, 2008). The latter allows for low storage percentages of approximately 20 wt%.

To enhance PHA accumulation from saline wastewater the implementation of a settling stage after the feast phase in the enrichment unit was assessed. An increase from 40 to 60% of maximum PHA accumulation was observed, along with an effect on the co-polymer composition, and HV proportion rose from 17 to 30% (Argiz *et al.*, 2020).

The last step of the process comprises the subsequent accumulation step where the MMC operates under nutrient limiting conditions to restrict biomass growth while the biopolymer accumulation is maximized. The maximal accumulation capacity of the biomass depends on the used substrate and the operational conditions imposed. For this reason, an adequate stream to be used as substrate requires careful selection to achieve, both, the correctly enriched culture and the large PHA accumulation yields (Rodriguez-Perez *et al.*, 2018).

Examples of mixtures of VFA produced from saline waste streams potentially appropriated for PHA storage are provided in Table 5.1. Their use diminishes PHA production costs by 50% (Jiang *et al.*, 2016) as the cost of the substrate represents 40% of the total PHA-production cost, according to Choi and Lee (1997), and non-sterile operational conditions are acceptable.

The produced PHA can be used as bioplastics, as feedstock for biofuel production and the production of chemicals like acrylate and propene (Gao *et al.*, 2011), or as a carbon source for denitrification (Santorio *et al.*, 2019).

With regard to the PHA, its sale price depends on the monomer composition and it usually varies in a range of 2.5–5.0 €/kg which is much lower than the price in the last decade (10–12 €/kg) (Valentino *et al.*, 2017). This price decreases with time as the production processes are improving, although there is still a long way to run to the meet the price of 1.0 €/kg corresponding to the petroleum-based plastics. For this reason, exploring the use of cheap organic carbon sources together with efficient and environmentally friendly processes is needed.

5.2.1.2.3 Triacylglycerides (TAG)

Triacylglycerides are storage compounds produced by eukaryotic microorganisms, plants, or yeasts, and to a less extent by bacteria (Álvarez, 2016). Oleaginous microorganisms can accumulate more than 20% of their weight as TAG. Like the case of PHA, these organisms use them as storage compounds for energy and carbon source to be used in periods of stress. The interest in TAG has increased lately due to their potential use in the production of biodiesel. Although they have been obtained from different feedstocks, like glycerol or agro-industrial residues (molasses, whey, olive mill), scarce information is available for their production from saline substrates. Fra-Vázquez *et al.* (2018) produced TAG from crude glycerol containing 14.5 g Na^+/L.

Operational conditions such as carbon to nitrogen ratio, oxygen concentration, and substrate addition strategy affect the lipid accumulation (Kosa and Ragauskas, 2011).

5.2.2 Nutrient recovery

Many types of salt-containing wastewater also contain large concentrations of nutrients, nitrogen and phosphorus, which aside from being removed can be recovered (Guerreiro *et al.*, 2020). In particular, phosphorus is obtained from rocks located in certain places in the world. It is a limited resource and, for this reason, its obtainment from other sources is under exploration. Among the most commonly applied processes to recover phosphorus from wastewater are those physical-chemical methods based on precipitation. Some of them are based on orthophosphate precipitation with lime, aluminium, iron, or magnesium salts. Among the latter, the precipitation of struvite ($MgNH_4PO_4 \cdot 6\ H_2O$), which contains nitrogen and phosphorus in the molecule, is an interesting option. This product has fertilizer potential (Ghosh *et al.*, 1996) and its commercialization is likely to increase due to the recent changes in the European Union regulations on fertilizing products. The new regulations consider struvite produced from wastewater, if it accomplishes certain composition requirements, as one of these products (Regulation (EU) 2019/1009).

The struvite precipitation process from wastewater requires the addition of a magnesium (Mg) source (magnesium hydroxide, magnesium chloride or bitter (Crutchik and Garrido, 2011)). The cost of these compounds affects the final cost of the produced struvite. To reduce the struvite production costs, seawater that contains over 1 g Mg/L was explored as a source of Mg (Crutchik and Garrido, 2011; Matsumiya *et al.*, 2000). Matsumiya *et al.* (2000) tested seawater as a Mg source to precipitate struvite from urban wastewater and observed that large Mg/P ratios (over 1.5) enhanced P removal while the increased ion concentration due to seawater did not affect the treatment process. When seawater addition was explored to crystallize struvite from fish canning wastewater its dosage and ammonium to phosphate ratio (molar ratio of 4) were identified as relevant factors affecting quality and efficiency, respectively (Crutchik and Garrido, 2011).

Some authors also claim that nutrients can be recovered by biomass production in microalgae systems like those based on *Spirulina platensis* (Zhou *et al.*, 2017).

In other cases, the potential of the biosolids produced from treatment of industrial wastewater to be used as fertilizers has been explored and is increasing in application. As an example, the sludge obtained from the physical-chemical treatment (based on the addition of 75 g $CaCO_3/L + 1$ g $FeCl_3/L$) of cheese manufacturing wastewater presented pH values between 5.5 and 8.5 and EC values below 3 000 $\mu S/cm$, and as a consequence was suitable for agricultural valorization (Guerreiro *et al.*, 2020).

5.2.3 Salts recovery

Apart from organic matter and nutrients, salts also need to be removed from wastewater except in some cases where it is treated to be discharged to the sea. Different technologies have been tested to remove salts from wastewater (Kim, 2011). Some of them have already been applied for seawater desalination or brine concentration.

Salts present in different types of liquid flows like seawater treated for desalination, oil and gas production processes, and the food processing industry when separation produces brines. In all these cases the concept of zero liquid discharge (ZLD) is considered to guarantee the appropriate management of the produced effluents to obtain high-quality freshwater without the necessity to discharge a liquid flow. This approach includes conventional wastewater treatment systems based on physical-chemical (clarification, precipitation, coagulation-flocculation and filtration) and biological processes, membrane technologies (micro, ultra and nanofiltration and reverse osmosis), thermal technologies for brine treatment (evaporation) and crystallization.

The train of technologies to apply is defined depending on the characteristics of the wastewater (flow and composition) and the requirements of the produced freshwater, etc. (Kim, 2011). They aim to recover salts and produce reusable wastewater.

Technologies for salt separation are based on thermal treatments (i.e. evaporation, thermal crystallization, spray drying), membrane technologies (i.e. reverse osmosis, nanofiltration, electrodialysis), etc.

Concentrated brine is obtained from the thermal crystallization of highly saline water (100–200 g TS/L) by removing the water (Chen *et al.*, 2018a). Salts like magnesium hydroxide, sodium chloride, calcium carbonate, sodium sulphate, calcium chloride, etc. can be recovered from brine obtained from the application of nanofiltration and reverse osmosis systems which help to improve the economics of the separation processes (Bazedi *et al.*, 2013; Kim, 2011).

In the case of the dairy industry, with salinities close to seawater, depending on the product the salt might be at high concentrations, as is the case in cheese production where sodium chloride salts are applied to the product to reduce water activity during the curd (Chen *et al.*, 2018a).

The energy consumption associated with the different commercial technologies applied to the treatment of saline wastewater for salt separation is widely variable and increases exponentially with increasing salinity concentration of the processed stream; however, treatment costs do not experience the same tendency (Chen *et al.*, 2018a). As an example, Kim (2011) estimated that the production of $CaCO_3$, NaCl and $MgSO_4 \cdot 7\ H_2O$ by nanofiltration or membrane crystallization costs \$18/t, while using the Dow chemical process the cost of magnesium production was of \$14,146/t. Magnesium is very expensive. Thus, whether the applied processes are viable or not relies on the sale price of the recovered product.

5.2.4 Energy recovery

Biogas is produced during anaerobic digestion of varying types of wastewater containing biodegradable organic compounds and, depending on its quality (presence of impurities like H_2O, CO_2, H_2S, NH_3, etc.) and CH_4 percentage, may be utilized to generate heat or electricity on-site, sold to a local natural gas provider, or used as fuel for vehicles. Biogas with impurities needs to be submitted to different pre-treatments to remove those compounds that are not wanted for a specific purpose depending on the established requirements.

Methane obtainment, from saline wastewater has been previously discussed in Chapter 3. Here an analysis from the point of view of the quality of the produced biogas is performed (Table 5.3).

Depending on the composition of the wastewater the produced biogas presents different characteristics. Methane percentage in the produced biogas when saline wastewater, such as that from fish canneries, is treated ranges between 45–80% (Xiao and Roberts, 2010).

The simultaneous production of CH_4 and H_2 from mono-ethylene glycol (MEG) containing salt concentrations in the range of 10–35 g NaCl/L has also been evaluated by Ali *et al.* (2019). These authors observed that at the same OLR the yield of methane and hydrogen slightly increased with the rise of the salt concentration from 15 to 35 g NaCl/L. The removed OLR also augmented from 70 to 80% with the salinity increase which was attributed to the biomass acclimation to the operational conditions.

When saline wastewater is digested anaerobically, apart from the inhibitory effect of the salts on the methanogenic activity, the formation of undesired gaseous compounds, such as hydrogen sulphide (H_2S), also takes place. For this reason, the quality, and consequently the usefulness, of the produced biogas

Table 5.3 Quality of the biogas produced from the anaerobic treatment of saline wastewater and post-treatment required for valorization purposes. All experiments were performed at 35–37°C.

Industrial Process	Salinity (g/L)	Experiment (System Volume, L)	OLR kg COD/($m^3 \cdot d$)	Biogas Production (m^3 Biogas/kg COD Removed)	m^3 CH_4/(m^3 Wastewater · d)	Methane Content (%)	Sulphide Content (%)	Reference
Fish canning	5.05–12.63 g Na^+/L 0.59–2.7 g SO_4^{2-}/L	Full-scale (15,000)	3	NA	NA	NA	4.25	Omil *et al.* (1995)
Molasses	20,000–90 000*	Laboratory (1.7)	10	0.349 m^3 CH_4/kg COD_{rem}	331	29	NA	Scoma *et al.* (2017)
Fish processing	20	Batch (0.25)	NA	0.49	0.2–0.5[a]	40–50	NA	Picos-Benítez *et al.* (2019)
Mono-ethylene glycol (MEG) 25°C	10 g NaCl/L	Laboratory (ISABR, 22.5)	0.64	0.165 m^3 CH_4/kg COD_{added} 0.102 m^3 H_2/kg COD_{added}	0.105 m^3 CH_4/($m^3 \cdot d$) 0.065 m^3 H_2/($m^3 \cdot d$)	49.51 30.74 (H_2%)	19.75[b]	Ali *et al.* (2019)
Mono-ethylene glycol (MEG) 35°C[c]	15–25 g NaCl/L	Laboratory (ISABR, 22.5)	0.64	0.322 m^3 CH_4/kg COD_{added} 0.239 m^3 H_2/kg COD_{added}	0.206 m^3 CH_4/($m^3 \cdot d$) 0.153 m^3 H_2/($m^3 \cdot d$)	35.2 26.07 (H_2%)	38.73[b]	Ali *et al.* (2019)

*In electrical conductivity units (μS/cm).
[a]Litres of biogas/kg COD removed.
[b]Percentage corresponding to other gases (CO_2 and H_2S).
[c]Average values of experiments performed with terrestrial and marine inoculum. ISABR: immobilized sludge anaerobic baffled reactor.

diminishes. The H_2S generates problems of corrosion and cannot be directly supplied to boilers for energy production due to the generation of SO_x in the exhaust gases. For this reason, before biogas utilization the removal of this compound is generally required.

As an example, Omil *et al.* (1995) measured concentrations of H_2S of 4.25% in the biogas at pH values of 7.25 and concentrations of 500 mg S/L of sulphide dissolved in the liquid phase, when the ratio of COD:SO_4^{2-} was 7 g/g.

Depending on the final use of the biogas and its H_2S concentration, a treatment to remove this compound could be required. The recommended percentage of H_2S in the produced biogas to be used for direct combustion is of 0.02 to 0.05% w/w (Khoshnevisan *et al.*, 2017). This removal generally takes place by air/oxygen dosing of the biogas, iron chloride addition to the digester slurry, or adsorption on iron oxide pellets or liquids, which are relatively expensive technologies due to the chemical products employed and the management of the produced chemical waste. More recent developments are based on bio scrubbers or biotrickling filters (Khoshnevisan *et al.*, 2017) where microorganisms are responsible for the H_2S removal.

Furthermore, water must be removed too in all cases as in combination with H_2S it forms acidic compounds which are corrosive. Water, foam and dust are removed simultaneously via condensation, drying by cooling, adsorption to silica or using glycol when the gas is to be applied as natural gas or vehicle fuel. In the case of small-sized AD plants, moisture or water traps are applied.

Also, the CO_2 needs to be removed, as it is a non-combustible compound, when it is going to be used for electricity production in some gas engines or injected in the natural gas network. When biogas is used for direct heating in gas boilers its presence is not a problem. Finally, dust is also a problematic pollutant which must be removed.

5.3 WATER REUSE POTENTIAL

As the volume of water accounts for the largest fraction in the wastewater composition, one of the most common approaches for reuse of reclaimed water relies on the removal, recovery or separation of the pollutants to generate an effluent with the required quality for valorization. Wastewater from both urban and industrial origin can be treated for reuse. When wastewater contains salts, the treatments indicated in Section 5.2.3 are used and a fraction of brine is produced simultaneously to the reclaimed wastewater. Of course, apart from the salts also pathogens, toxic compounds (as micropollutants) and eventually nutrients have to be removed to reach required limits.

5.3.1 Reuse alternatives

For reuse purposes the salt present in the wastewater is a limiting parameter as it is not removed by the treatment processes of wastewater treatment plants (WWTPs).

Therefore, when it exceeds the concentration limit for reuse the implementation of a post-treatment to remove it from the final effluent is required. In these cases, membrane treatments like those based on reverse osmosis (RO) are commonly applied.

When appropriate conditions are achieved, reclaimed water, produced after secondary or tertiary treatments in wastewater treatment plants, can be used for different purposes such as irrigation, groundwater recharge, recreational uses, urban/domestic uses, indirect potable reuse, industrial uses, environmental uses, etc. (Table 5.4). The utilization of reclaimed water as a resource increases the water supply by 17–30% in small countries like Israel and Malta (Hochstrat *et al.*, 2008).

Compared to the obtainment of water from seawater desalination plants for non-potable applications, the use of reclaimed wastewater is less expensive as its production requires less energy consumption (Meneses *et al.*, 2010). To reuse reclaimed wastewater, apart from the implementation of appropriate treatment processes, building of the distribution system is required. Thus, costs associated with elements like pipes, tanks, reservoir pumps, etc. must be considered to evaluate the feasibility of this option.

The reuse activities differ depending on the region of the world and as an example agricultural irrigation is the main reuse option in Europe while in Australia industrial reuse is the most common (Hochstrat *et al.*, 2008). The most common uses for reclaimed wastewater are agricultural irrigation, groundwater recharge, and industrial uses, and, for this reason, they will be discussed in the following sections.

5.3.1.1 Agricultural irrigation

Agricultural irrigation is the most frequent use for regenerated wastewater being the activity that consumes by far the most water. Planned irrigation with reclaimed wastewater in Spain has been a frequent practice since the second half of the last century (Jodar-Abellan *et al.*, 2019) and it is also widespread all over the world (Cherif *et al.*, 2013; Darwish *et al.*, 2015). In countries such as Qatar 92% of groundwater abstraction is used for agricultural purposes (Darwish *et al.*, 2015). Irrigation with saline water exerts negative effects on vegetables such as decreasing water uptake into the plant, causing toxicity to the plant, and altering soil structure and permeability (Kathijotes and Panayiotou, 2013). However, it was reported that the irrigation of grapefruit trees, using drip irrigation, for four years with saline reclaimed wastewater, from a WWTP including an UV tertiary treatment, did not significantly affect the plant production or fruit quality compared to the use of river irrigation water (Pedrero *et al.*, 2013). These authors report on the possible risk occurrence of soil salinization due to medium and long-term use of boron and salt-containing wastewater.

Table 5.4 Reuse examples of reclaimed saline wastewater from urban and industrial origin.

Source of Reclaimed Wastewater	Scale	Salinity of Reclaimed Water (g/L)	Treatment for Recovery	Type of Valorization	Reference
Agricultural irrigation					
Urban wastewater	Full-scale	3,000–4,000 μS/cm	Tertiary treatment (UV)	Drip irrigation of grapefruit trees	Pedrero *et al.* (2013)
Urban wastewater	Full-scale	2,311 2,311 57.09 μS/cm	CFF + UV CFF + UF CFF + UF + RO	Except for fresh food All agricultural uses All agricultural uses	Melgarejo *et al.* (2016)
Urban wastewater	Full-scale	2.7 μS/cm	UF + UV (when needed)	Tomato fruits and broccoli heads	Libutti *et al.* (2018)
Fish processing	Laboratory	0–25 g NaCl/L	Biological aerobic treatment	Liquid fertilizer	Ching and Redzwan (2017)
Cheese manufacturing	Laboratory	4,880 μS/cm	Physical-chemical $CaCO_3 + FeCl_3$, $FeSO_4$ or $Ca(OH)_2$	Potential agricultural irrigation	Guerreiro *et al.* (2020)
Petrochemical oily ww	Laboratory	27,345–38,350 μS/cm	Forward Osmosis (FO)	Direct fertigation	Sahebi *et al.* (2020)

(*Continued*)

Table 5.4 Reuse examples of reclaimed saline wastewater from urban and industrial origin (*Continued*).

Source of Reclaimed Wastewater	Scale	Salinity of Reclaimed Water (g/L)	Treatment for Recovery	Type of Valorization	Reference
Aquifer recharge					
Urban wastewater	Full-scale	1.8–5.4 g salinity/L	Low-load activated sludge plus 3 infiltration ponds	Protect the aquifer against marine intrusions	Cherif *et al.* (2013)
Urban uses					
Urban wastewater	Full-scale	2,311 2,311 57.09 µS/cm	CFF + UV CFF + UF CFF + UF + RO	Residential and urban services Urban services Urban services	Melgarejo *et al.* (2016)
Industrial application					
Power plant effluents	Full-scale Full-scale	975–3,475 µS/cm	Tubular reverse osmosis Electrodialysis reversal	Plant process Plant process	Schutte *et al.* (1987)
Tannery	Full-scale	140–740	RO	Plant process	Ranganathan and Kabadgi (2011)
Fish canning wastewater	Laboratory	4,730–24,800 µS/cm	Tertiary treatment Filtration, RO, UV	Plant process	Cristóvão *et al.* (2015)

CFF = Coagulation + flocculation + filtration.

Research studies performed to evaluate the use of reclaimed wastewater, from a WWTP with seawater intrusions (3.75 mS/cm), for irrigation indicated that the levels of salts, EC and the sodium absorption ratio (SAR) increased in all layers of the soil while this did not happen when using freshwater (Kathijotes and Panayiotou, 2013). According to these authors soils irrigated with wastewater become saline with time, thus the quality of effluents from WWTPs must be carefully monitored.

Reclaimed water can be used not only to supply water but also for fertigation (application of fertilizer with drip irrigation) to provide nutrients (nitrogen and phosphorus and potassium) to the crops. Therefore, tertiary treatment for nutrient removal in the WWTP is not needed and wastewater can go from the secondary treatment directly to the disinfection step. Salt removing technologies like RO, which are expensive, are not recommended either as they remove nutrients as well. So, salts remain in the wastewater. Although reclaimed water appropriate for this use is obtained, some organizational and technological limitations, associated with the stringent normative requirements, need to be solved to extensively implement this valorization option (Chojnacka *et al.*, 2020). According to these authors the hypersalinity and sodium toxicity risks are important parameters of the wastewater and, associated with them, sodium content should be monitored to guarantee the effective removal of salt from the plant root zones. Melgarejo *et al.* (2016) evaluated three treatment tertiary treatments for wastewater with EC values of 2,300 μS/cm including UF (ultrafiltration) and RO (Table 5.4) and determined that irrigation was feasible for the three cases.

Although the most frequently used reclaimed wastewater for irrigation and fertigation is that from urban origin, industrial wastewater has also been used for these purposes.

The reclaimed water from the agro-industry appears to be an option for irrigation of vegetable crops. As an example, Libutti *et al.* (2018) reused 10 000 m^3/ha of water and successfully subsequently irrigated tomato and broccoli crops during a period of 1.5 years and managed to save 6 000 m^3/ha per year. These authors successfully applied the water coming from the tertiary treatment of the WWTP comprising a UF and subsequent UV unit with an EC of 2.7 μS/cm.

Guerreiro *et al.* (2020) succeeded in obtaining water suitable for agriculture irrigation by applying a coagulation-flocculation post-treatment to cheese manufacturing wastewater, containing organic matter and nutrients, with an EC of 4,880 μS/cm.

In another study Ching and Redzwan (2017) already pointed out the opportunity for reuse of treated fish processing wastewater containing different salt concentrations as liquid fertilizer, containing nitrogen, proteins and phosphorus, in Malaysia; however, they highlighted the necessity of further studies to evaluate its feasibility as a fertilizer and potential effect on irrigated soil, plants and yields of crops.

More recently the use of forward osmosis (FO) has been tested for the treatment of petrochemical oily wastewater containing as average concentrations 1,020 mg COD/L, 118 mg BOD_5/L (BOD_5, biological oxygen demand measured after 5 days), 64 mg TSS/L (TSS, total suspended solids) and 32,848 μS/cm (as sodium salts). Potassium chloride was used as the draw solution for the FO due to its capability for pure water draw (Phuntsho *et al.*, 2011), and its direct use for fertigation was evaluated to develop a fertilizer drawn forward osmosis (FDFO) (Sahebi *et al.*, 2020). With this technology 40% of the water volume was reclaimed without the need to apply any hydraulic pressure.

5.3.1.2 Groundwater recharge

More than half of the biggest cities in the world, those with more than 10 million inhabitants, depend on groundwater extraction from aquifers which are submitted to abstraction of large water volumes (Werner *et al.*, 2013). These authors identified the groundwater extraction and population as factors affecting the groundwater depletion, and identified China, Spain, Mexico, the USA, and Pakistan as nations with the largest groundwater depletion. In countries like Israel the effluent from the WWTPs is injected into the groundwater to replenish those bodies submitted to freshwater extraction. Although in general this action is taken to reduce salinity from the groundwater by recharging the aquifers with non-saline treated municipal wastewater (Horriche and Benabdallah, 2020).

Groundwater can be recharged by surface spreading or direct injection to the aquifer (Asano and Cotruvo, 2004). In the first case effluents from WWTPs percolate from the discharge point to the aquifer and the water quality improves due to the filtration process. In the second option water with high quality is injected deeply into the aquifer with no intermediate percolating step.

Thus, recharge of aquifers with reclaimed wastewater needs to consider the characteristics of the groundwater. For example, aquifers located on the coastline can be replenished with low quality wastewater just to prevent seawater intrusions due to over abstraction practices (Franks *et al.*, 2007; Hussain *et al.*, 2017), while those placed inland where potable water can be extracted require high quality water.

Uses of water from aquifers, such as potable water, animal drinking and agricultural irrigation, define its required quality.

Cherif *et al.* (2013) evaluated the replenishment of an aquifer with a salinity of 8.5 g/L, to act as a barrier against marine intrusions, with treated water with a salinity of 1.8–5.4 g/L, depending on the season, using three infiltration basins to retain suspended solids, oxidize organic matter and nitrogen, and remove pathogens. In this study samples of groundwater were collected for three years and a significant decrease of salinity, to 0.5–1.5 g/L, was observed in the area close to the recharge plant, however contamination of nitrate and pathogens was not yet solved.

5.3.1.3 *Industrial reuse*

Among the reuse alternatives of industrial effluents are reutilization in the industrial production process, which is quite a rare option, or use in refrigeration towers, as boiler fed water and for external area cleaning or irrigation. In this way industrial facilities manage to decrease the water consumption of the production process which reduces operational and waste management costs. Although it is not a thoroughly explored option, some studies have evaluated treated industrial effluent for reuse (Table 5.4).

Schutte *et al.* (1987) managed to recycle, from a full-scale power plant, approximately 70% of the wastewater using tubular reverse osmosis (TRO) to return it to the cooling water loop, and 90% using electrodialysis reversal (EDR) to be reused in the process for different purposes including as feed to the boiler feedwater demineralization plant.

Ranganathan and Kabadgi (2011) monitored the performance of RO systems in operation in five different tannery facilities producing wastewater characterized by EC values of 6 000–15 900 μS/cm. The resulting permeate from the RO had an EC value of 140–740 μS/cm and approximately 70–85% of the wastewater from the industrial operation was able to be recycled. In this case, the RO rejects were evaporated and landfilled in a secure site.

Reuse of treated wastewater has been evaluated not only in the chemical industry but also in the agro-food industry. In this sense, Cristóvão *et al.* (2015) evaluated the production of water of sufficient quality to be used in a fish canning factory after the treatment of the wastewater generated by the facility. However, they did not specify for which use. The wastewater from the fish cannery was submitted to a tertiary treatment comprising filtration, RO and UV steps resulting in a decrease in the EC value of 99.1%.

5.4 CONCLUSIONS

Options for wastewater valorization are diverse and include product obtainment and water reclamation. Many of them are technically, economically, and environmentally feasible but their implementation at industrial scale is limited by the applicable legislation and public engagement.

Water reclamation is increasing in importance in those countries with water scarcity in particular to fulfil agricultural irrigation needs. This kind of reuse relies on the existence of a WWTP with appropriate treatment to provide an effluent with sufficient quality for this purpose. This implies that countries with a lack of freshwater resources, like Qatar (Darwish *et al.*, 2015), are increasing the volume of wastewater submitted to treatment and consequently the number of WWTPs. Costs associated with the construction of the reclaimed water distribution system must be considered.

REFERENCES

Ajao V., Millah S., Gagliano M. C., Bruning H., Rijnaarts H. and Temmink H. (2019). Valorization of glycerol/ethanol-rich wastewater to bioflocculants: recovery, properties, and performance. *Journal of Hazardous Materials*, **375**, 273–280.

Ali M., Elreedy A., Ibrahim M. G., Fujii M. and Tawfik A. (2019). Hydrogen and methane bio-production and microbial community dynamics in a multi-phase anaerobic reactor treating saline industrial wastewater. *Energy Conversion and Management*, **186**, 1–14.

Alvarez H. M. (2016). Triacylglycerol and wax ester-accumulating machinery in prokaryotes. *Biochimie*, **120**, 28–39.

Argiz L., Fra-Vázquez A., Val del Río A. and Mosquera-Corral A. (2020). Optimization of an enriched mixed culture to increase PHA accumulation using industrial saline complex wastewater as a substrate. *Chemosphere*, **247**, 125873.

Asano T. and Cotruvo J. A. (2004). Groundwater recharge with reclaimed municipal wastewater: health and regulatory considerations. *Water Research*, **38**, 1941–1951.

Atasoy M., Owusu-Agyeman I., Plaza E. and Cetecioglu Z. (2019). Bio-based volatile fatty acid production and recovery from waste streams: current status and future challenges. *Bioresource Technology*, **268**, 773–786.

Bazedi G. A., Ettouney R. S., Tewfik S. R., Sorour M. H. and El-Rifai M. A. (2013). Salt recovery from brine generated by large-scale seawater desalination plants. *Desalination and Water Treatment*, **52**, 4689–4697.

Bengtsson S., Hallquist J., Werker A. and Welander T. (2008). Acidogenic fermentation of industrial wastewaters: effects of chemostat retention time and pH on volatile fatty acids production. *Biochemical Engineering Journal*, **40**, 492–499.

Cervantes F. J. and Pavlostathis S. G. (2006). Strategies for industrial water pollution control (Chapter 1). In: Advanced Biological Treatment Processes for Industrial Wastewater, F. J. Cervantes, S. G. Pavlostathis and A. C. van Haandel (eds), IWA Publishing, London, UK, 1–15.

Chen Y., Jiang S., Yuan H., Zhou Q. and Gu G. (2007). Hydrolysis and acidification of waste activated sludge at different pHs. *Water Research*, **41**(3), 683–689.

Chen G. Q., Gras S. L. and Kentish S. E. (2018a). Separation technologies for salty wastewater reduction in the dairy industry. *Separation and Purification Reviews*, **00**, 1–29.

Chen Y., He H., Liu H., Li H., Zeng G., Xia X. and Yang C. (2018b). Effect of salinity on removal performance and activated sludge characteristics in sequencing batch reactors. *Bioresource Technology*, **249**, 890–899.

Cherif S., El Ayni F., Jrad A. and Trabelsi-Ayadi M. (2013). Aquifer recharge by treated wastewaters: Korba case study (Tunisia). *Sustainable Sanitation Practice*, **14**, 41–48.

Ching Y. C. and Redzwan G. (2017). Biological treatment of fish processing saline wastewater for reuse as liquid fertilizer. *Sustainability*, **9**(1062), 1–26.

Choi J. I. and Lee S. Y. (1997). Process analysis and economic evaluation for poly (3-hydroxybutyrate) production by fermentation. *Bioprocess Engineering*, **17**(6), 335–342.

Chojnacka K., Witek-Krowiak A., Moustakas K., Skrzypezak D., Mikula K. and Loizidou M. (2020). A transition from conventional irrigation to fertigation with reclaimed wastewater: prospects and challenges. *Renewable and Sustainable Energy Reviews*, **130**, 109959.

Cristóvão R. O., Botelho C. M., Martins R. J. E., Loureiro J. M. and Boaventura R. A. R. (2015). Fish caning industry wastewater treatment for water reuse – a case study. *Journal of Cleaner Production*, **87**, 603–612.

Crutchik D. and Garrido J. M. (2011). Struvite crystallization versus amorphous magnesium and calcium phosphate precipitation during the treatment of a saline industrial wastewater. *Water Science and Technology*, **64**(12), 2460–2467.

Crutchik D., Morales N., Vazquez-Padin J. R. and Garrido J. M. (2017). Enhancement of struvite pellets crystallization in a full-scale plant using an industrial grade magnesium product. *Water Science and Technology*, **75**(3–4), 609–618.

Dahiya S., Sarkar O., Swamy Y. V. and Mohan S. V. (2015). Acidogenic fermentation of food waste for volatile fatty acid production with co-generation of biohydrogen. *Bioresource Technology*, **182**, 103–113.

Darwish M., Abdulrahim H., Mabrouk A. N., Hassan A. and Shomar B. (2015). Reclaimed wastewater for agriculture irrigation in Qatar. *Global Journal of Agricultural Research and Reviews*, **3**(1), 106–120.

De Vrieze J., Christiaens M. E. R., Walraedt D., Devooght A., Ijaz U. Z. and Boon N. (2017). Microbial community redundancy in anaerobic digestion drives process recovery after salinity exposure. *Water Research*, **111**, 109–117.

Franks R., Bartels C. R. and Andes K. (2007). Implementing energy saving RO technology in large scale wastewater treatment plants. IDA World Congress-Maspalomas, Gran Canaria, 21–26 October.

Fra-Vázquez A., Pedrouso A., Palmeiro-Sánchez T., Moralejo-Gárate H. and Mosquera-Corral A. (2018). Feasible microbial accumulation of triacylglycerides from crude glycerol. *Journal of Chemical Technology and Biotechnology*, **93**, 2644–2651.

Fra-Vázquez A., Pedrouso A., Val del Río A. and Mosquera-Corral A. (2020). Volatile fatty acid production from saline cooked mussel processing wastewater at low pH. *Science of the Total Environment*, **732**, 139337.

Gao X., Chen J. C., Wu Q. and Chen G. Q. (2011). Polyhydroxyalkanoates as a source of chemicals, polymers, and biofuels. *Current Opinion in Biotechnology*, **22**, 768–774.

Ghosh G. K., Mohan K. S. and Sarkar A. K. (1996). Characterization of soil-fertilizer P reaction products and their evaluation as sources of P for gram (*Cicer arietinum* L.). *Nutrient Cycling in Agroecosystems*, **46**(1), 71–79.

Guerreiro R. C. S., Jerónimo E., Luz S., Pinheiro H. M. and Prazeres A. R. (2020). Cheese manufacturing wastewater treatment by combined physicochemical processes for reuse and fertilizer production. *Journal of Environmental Management*, **264**, 110470.

Guest J. S., Skerlos S. J., Barnard J. L., Beck M. B., Daigger G. T., Hilger H., Jackson S. J., Karvazy K., Kelly L., Macpherson L., Mihelcic J. R., Pramanik A., Raskin L., van Loosdrecht M. C. M., Yeh D. and Love N. G. (2009). A new planning and design paradigm to achieve sustainable resource recovery from wastewater. *Environmental Science & Technology*, **43**(16), 5.

Hochstrat R., Wintgens T. and Melin T. (2008). Development of integrated water reuse strategies. *Desalination*, **218**, 208–217.

Horriche F. J. and Benabdallah S. (2020). Assessing aquifer level and salinity for a managed artificial recharge site using reclaimed water. *Water*, **12**(2), 341–352.

Hussain M. S., Javadi A. A. and Sherif M. M. (2017). Cost-efficient management of coastal aquifers by recharging with treated wastewater. *International Journal of Hydrology*, **1** (6), 183–185.

Ismail S. B., de La Parra C. J., Temmink H. and van Lier J. B. (2010). Extracellular polymeric substances (EPS) in upflow anaerobic sludge blanket (UASB) reactors operated under high salinity conditions. *Water Research*, **44**, 1909–1917.

Jiang G., Hill D. J., Kowalczuk M., Johnston B., Adamus G., Irorere V. and Radecka I. (2016). Carbon sources for polyhydroxyalkanoates and an integrated biorefinery. *International Journal of Molecular Sciences*, **17**(7), 1157–1178.

Jodar-Abellan A., López-Ortiz M. I. and Melgarejo-Moreno J. (2019). Wastewater treatment and water reuse in Spain. Current situation and perspectives. *Water*, **11**(8), 1551–1574.

Johnson K., Jiang Y., Kleerebezem R., Muyzer G. and van Loosdrecht M. C. M. (2009). Enrichment of a mixed bacterial culture with a high polyhydroxyalkanoate storage capacity. *Biomacromolecules*, **10**(4), 670–676.

Kathijotes N. and Panayiotou C. (2013). Wastewater reuse for irrigation and seawater intrusion: evaluation of salinity effects on soils in Cyprus. *Journal of Water Reuse and Desalination*, **3**(4), 392–401.

Khoshnevisan B., Tsapekos P., Alfaro N., Díaz N., Fdz-Polanco M., Rafiee S. and Angelidaki I. (2017). A review on prospects and challenges of biological H_2S removal from biogas with focus on biotrickling filtration and microaerobic desulfurization. *Biofuel Research Journal*, **16**, 741–750.

Kim D. H. (2011). A review of desalting process technique and economic analysis of the recovery of salts from retentates. *Desalination*, **270**, 1–8.

Kleerebezem R., Joosse B., Rozendal R. and Van Loosdrecht M. C. M. (2015). Anaerobic digestion without biogas? *Reviews in Environmental Science and Bio/Technology*, **14** (4), 787–801.

Kosa M. and Ragauskas A. J. (2011). Lipids from heterotrophic microbes: advances in metabolism research. *Trends in Biotechnology*, **29**(2), 53–61.

Lefebvre O. and Moletta R. (2006). Treatment of organic pollution in industrial saline wastewater: a literature review. *Water Research*, **40**, 3671–3682.

Libutti A., Gatta G., Gagliardi A., Vergine P., Pollice A., Beneduce L., Disciglio G. and Tarantino E. (2018). Agro-industrial wastewater reuse for irrigation of a vegetable crop succession under Mediterranean conditions. *Agricultural Water Management*, **196**, 1–14.

Lin Y., de Kreuk M., van Loosdrecht M. C. M. and Adin A. (2010). Characterization of alginate-like exopolysaccharides isolated from aerobic granular sludge in pilot-plant. *Water Research*, **44**(11), 3355–3364.

Liu X., Dai J., Wu D., Jiang F., Chen G., Chui H.-K. and van Loosdrecht M. C. M. (2016). Sustainable application of a novel water cycle using seawater for toilet flushing. *Engineering*, **2**, 460–469.

Matsumiya Y., Yamasita T. and Nawamura Y. (2000). Phosphorus removal from sidestreams by crystallization of magnesiumammonium-phosphate using seawater. *Journal of the Chartered Institution of Water and Environmental Management*, **14**, 291–296.

Melgarejo J., Prats D., Molina A. and Trapote A. (2016). A case study of urban wastewater reclamation in Spain: comparison of water quality produced by using alternative processes and related costs. *Journal of Water Reuse and Desalination*, **6**(1), 72–81.

Meneses M., Pasqualino J. C. and Castells F. (2010). Environmental assessment of urban wastewater reuse: treatment alternatives and applications. *Chemosphere*, **81**, 266–272.

Moralejo-Gárate H., Mar'atusalihat E., Kleerebezem R. and van Loosdrecht M. M. (2011). Microbial community engineering for biopolymer production from glycerol. *Applied Microbiology and Biotechnology*, **92**(3), 631–639.

Mosquera-Corral A., Carvalho G., Fra-Vázquez A., Ntaikou I., Oleskowicz-Popiel P., Palmeiro-Sánchez T., Reis M. A. and Suárez-Ojeda M. E. (2017). Recovery of organic added value products from wastewater (Chapter 18). In: Innovative Wastewater Treatment and Resource Recovery Technologies, J. M. Lema and S. Suárez (eds), IWA Publishing, London, UK, 399–420.

Omil F., Méndez R. and Lema J. M. (1995). Anaerobic treatment of saline wastewaters under high sulphide and ammonia content. *Bioresource Technology*, **54**, 269–278.

Palmeiro-Sanchez T., Fra-Vazquez A., Rey-Martinez N., Campos J. L. and Mosquera-Corral A. (2016a). Transient concentrations of NaCl affect the PHA accumulation in mixed microbial culture. *Journal of Hazardous Materials*, **306**, 332–339.

Palmeiro-Sánchez T., Oliveira C. S. S., Gouveia A. R., Noronha J. P., Ramos A. M., Mosquera-Corral A. and Reis M. A. M. (2016b). NaCl presence and purification affect the properties of mixed culture PHAs. *European Polymer Journal*, **85**, 256–265.

Pedrero F., Alarcón J. J., Nicolás E. and Mounzer O. (2013). Influence of irrigation with saline reclaimed water on young grapfruits. *Desalination and Water Treatment*, **51**, 2488. .2496.

Phuntsho S., Shon J. K., Hong S., Lee S. and Vigneswaran S. (2011). A novel low energy fertilizer driven forward osmosis desalination for direct fertigation: evaluating the performance of fertilizer draw solutions. *Journal of Membrane Science*, **375**, 172–181.

Picos-Benítez A. R., Peralta-Hernández J. M., López-Hincapié J. D. and Rodríguez-García A. (2019). Biogas production from saline wastewater of the evisceration process of the fish processing industry. *Journal of Water Process Engineering*, **32**, 100933.

Pramanik A., Mitra A., Arumugam M., Bhattacharyya A., Sadhukhan S., Ray A., Haldar S., Mukhopadhyay U. K. and Mukherjee J. (2012). Utilization of vinasse for the production of polyhydroxybutyrate by *Haloarcula marismortui*. *Folia Microbiologica*, **57**, 71–79.

Puyol D., Batstone D. J., Hülsen T., Astals S., Peces M. and Krömer J. O. (2016). Resource Recovery from Wastewater by Biological Technologies: opportunities, Challenges, and Prospects. *Frontiers in Microbiology*, **7**, 2106.

Quillaguamán J., Guzmán H., Van-Thuoc D. and Hatti-Kaul R. (2010). Synthesis and production of polyhydroxyalkanoates by halophiles: current potential and future prospects. *Applied Microbial Biotechnology*, **85**(6), 1687–1696.

Rajeshwari K. V., Balakrishnan M., Kansal A., Lata K. and Kishore V. V. N. (2000). State-of-the-art of anaerobic digestion technology for industrial wastewater treatment. *Renewable and Sustainable Energy Reviews*, **4**(2), 135–156.

Ranganathan K. and Kabadgi S. D. (2011). Studies on the feasibility of reverse osmosis (membrane) technology for treatment of tannery wastewater. *Journal of Environmental Protection*, **2**, 37–46.

Regulation (EU) 2019/1009 of the European parliament and the Council of 5 June 2019 laying down rules on the making available on the market of EU fertilising products and amending Regulations (EC) No 1069/2009 and (EC) No 1107/2009 and repealing Regulation (EC) No 2003/2003 (OJ L 170, 25.6.2019, p. 1).

Reis M. A. M., Serafim L. S., Lemos P. C., Ramos A. M., Aguiar F. R. and van Loosdrecht M. C. M. (2003). Production of polyhydroxyalkanoates by mixed microbial cultures. *Bioprocess and Biosystems Engineering*, **25**(6), 377–385.

Rodriguez-Perez S., Serrano A., Pantión A. A. and Alonso-Fariñas B. (2018). Challenges of scaling-up PHA production from waste streams. A review. *Journal of Environmental Management*, **205**(Supplement C), 215–230.

Ruiken C. J., Breuer G., Klaversma E., Santiago T. and van Loosdrecht M. C. M. (2013). Sieving wastewater – Cellulose recovery, economic and energy evaluation. *Water Research*, **47**, 43–48.

Sahebi S., Sheikhi M., Ramavandi B., Ahmadi M., Zhao S., Adeleye A. S., Shabani Z. and Mohammadi T. (2020). Sustainable management of saline oily wastewater via forward osmosis using aquaporin membrane. *Process Safety and Environmental Protection*, **138**, 199–207.

Santorio S., Fra-Vazquez A., Val del Rio A. and Mosquera-Corral A. (2019). Potential of endogenous PHA as electron donor for denitrification. *Science of the Total Environment*, **695**, 133747.

Schutte C. F., Spencer T., Aspden J. D. and Hanekom D. (1987). Desalination and reuse of power plant effluents: frompilot plant to full scale application. *Desalination*, **67**, 255–269.

Scoma A., Coma M., Kerckhof F.-M., Boon N. and Rabaey K. (2017). Efficient molasses fermentation under high salinity by inocula of marine and terrestrial origin. *Biotechnology for Biofuels*, **10**(23), 1–17.

Serafim L. S., Lemos P. C., Albuquerque M. G. and Reis M. A. (2008). Strategies for PHA production by mixed cultures and renewable waste materials. *Applied Microbiology and Biotechnology*, **81**(4), 615–628.

Tan D., Xue Y. S., Aibaidula G. and Chen G. Q. (2011). Unsterile and continuous production of polyhydroxybutyrate by Halomonas TD01. *Bioresource Technology*, **102**, 8130–8136.

Thomson N., Roy I., Summers D. and Sivaniah E. (2010). In vitro production of polyhydroxyalkanoates: achievements and applications. *Journal of Chemical Technology and Biotechnology*, **85**(6), 760–767.

Valentino F., Morgan-Sagastume F., Campanari S., Villano M., Werker A. and Majone M. (2017). Carbon recovery from wastewater through bioconversion into biodegradable polymers. *New Biotechnology*, **37**(Part A), 9–23.

Venkata Mohan S., Nikhil G. N., Chiranjeevi P., Nagendranatha Reddy C., Rohit M. V., Kumar A. N. and Sarkar O. (2016). Waste biorefinery models towards sustainable circular bioeconomy: critical review and future perspectives. *Bioresource Technology*, **215**, 2–12.

Werner A. D., Xhang Q., Xue L., Smerdon B. D., Li X., Zhu X., Yu L. and Li L. (2013). An initial inventory and indexation of groundwater mega-depletion cases. *Water Resource Management*, **27**, 507–533.

Xiao Y. and Roberts D. J. (2010). A review of anaerobic treatment of saline wastewater. *Environmental Technology*, **31**(8–9), 1025–1043.

Xue W., Zeng Q., Lin S., Zan F., Hao T., Lin Y., van Loosdrecht M. C. M. and Chen G. (2019). Recovery of high-value and scarce resources from biological wastewater treatment: Sulfated polysaccharides. *Water Research*, **163**, 114889.

Yin J., Yu X., Wang K. and Shen D. (2016). Acidogenic fermentation of the main substrates of food waste to produce volatile fatty acids. *International Journal of Hydrogen Energy*, **41**(46), 21713–21720.

Zeng J., Gao J.-M., Chen Y.-P., Yan P., Dong Y., Shen Y., Guo J.-S., Zeng N. and Zhang P. (2016). Composition and aggregation of extracellular polymeric substances (EPS) in hyperhaline and municipal wastewater treatment plants. *Scientific Reports*, **6**, 26721.

Zhang B., Zhang L. L., Zhang S. C., Shi H. Z. and Cai W. M. (2005). The influence of pH on hydrolysis and acidogenesis of kitchen wastes in two-phase anaerobic digestion. *Environmental Technology*, **26**, 329–340.

Zhou W., Li Y., Gao Y. and Zhao H. (2017). Nutrients removal and recovery from saline wastewater by *Spirulina platesis*. *Bioresource Technology*, **245**, 10–17.

Zhou M., Yan B., Wong J. W. C. and Zhang Y. (2018). Enhanced volatile fatty acids production from anaerobic fermentation of food waste: a mini-review focusing on acidogenic metabolic pathways. *Bioresource Technology*, **248**, 68–78.

Chapter 6

Case study: Treatment of fish-canning effluents

ABSTRACT

The seafood processing industry represents an important sector in Spain in terms of benefits and employment. In the elaboration of seafood cans, significant amounts of wastewater are generated. The compliance of these facilities with the more stringent regulations implies the development of wastewater treatment processes able to cope with the variable composition of these effluents and the presence of salinity. In the present chapter the current situation regarding the fish-canning industry and the applied wastewater treatment systems are presented and discussed. Special importance is paid to those designing and operating wastewater treatment facilities. Available results from full-scale plants performance are provided here.

Keywords: Composition, discharge limit, fats, reuse

6.1 MOTIVATION

The fish and seafood industry is an important sector in Spain as it contributes 80% of the European production and makes Spain the top producing country of canned tuna in the EU (EUMOFA, 2019) and the second one in the world. It remains among the top five exporters in the world and was the biggest exporter from the EU in 2019. During the product processing, large amounts of wastewater are produced which need to be further treated before discharge to natural bodies. In many cases, wastewater produced contains salts and, after treatment, it is disposed of through discharge pipes to the sea or less frequently to sewage collectors. Nowadays,

Author(s): Anuska Mosquera Corral, Ángeles Val del Río and José Luis Campos Gómez
doi: 10.2166/9781789060645_0123

the new normative requirement with regard to effluent quality for discharge is becoming more stringent and is forcing the fish-canning sector to implement more efficient wastewater treatment processes.

Since the fish-processing industry needs to treat the generated wastewater, this sector is experiencing certain problems associated with the costs and performance of the treatment systems which affect the benefits of the companies. For these reasons, it is crucial to develop and implement the appropriate wastewater treatment systems able to cope with both the pollutant load and the saline conditions choosing, among the most efficient, an economic option. Furthermore, if treated water of sufficient quality is produced it can be valorized for reuse helping to save existing freshwater resources.

6.2 CHARACTERISTICS AND FLOWS OF FISH-CANNING EFFLUENTS

Over 90% of fish products are from marine origin, and for this reason seawater is widely used (Xiao & Roberts, 2010). Moreover, the different fish types are caught seasonally in their periods of production and when the fishing grounds open. Associated seawater consumption of 20 m^3 per tonne of fish captured is reported (Xiao & Roberts, 2010).

The fish canneries process the fish together with other seafood products depending on their availability and price of the raw product. During processing part of the raw fish, which is not transformed into the product, is diverted as a residue. This amount is approximately 50% of the fresh fish. Many of the process activities employ water as a utility. Moreover, the volumes of consumed water are very variable depending on the technologies used which many facilities have improved to reduce this consumption (Uttamangkabovorn *et al.*, 2005). Consequently, solid, and liquid streams are produced in the fish canning facility. The solid share comprises fractions of raw material wasted during the different processing steps and it is frequently valorized for fishmeal for animal feed production.

The liquid streams are generated in the different processing operations including defrosting, washing, cooking, and canning. The different steps of these processes generate wastewater streams with different characteristics and these steps will depend on the size of the facility. Depending on the step of the processing, the amount of wastewater produced can range between 0.2 m^3/ton of product, in the scaling step, and 5–11 m^3/ton during the filleting step (Ghaly *et al.*, 2013). In addition, the composition of this type of wastewater varies according to factors like the type of species (Table 6.1), sex, age, nutritional status, time of year and health. In general, fish composition comprises crude protein (58%) and fats (19%) which are released to the produced wastewater (Suvanich *et al.*, 2006).

Small factories differ from the large ones in the mechanization of the different steps that generally help to reduce the water consumption. Normally the

Table 6.1 compositions of different types of wastewater corresponding to different processes.

Wastewater Type	T (°C)	pH	tCOD (g/L)	sCOD (g/L)	TSS (g/L)	VSS (g/L)	NH_4^+-N (g/L)	PO_4^{-3}-P (g/L)	Prot. (g/L)	Fats (g/L)	Carbohyd. (g/L)	Cl^- (g/L)	SO_4^{2-} (g/L)	References
Tuna	85	5.7	50	44	2.7	2.6	0.2	0.5	14.8	–	–	94	–	Veiga (1994)
	–	6.96	5.55	–	1.57	–	0.44[a]	0.05	–	1.45	–	1.85	–	Achour (2000)
Tuna	45	–	34.5	–	4.0[b]	3.6[b]	–	–	77	12	11	14.0	–	Méndez (1992)
	100	–	18.5	–	1.4	1.2	–	–	22	4	74	13.0	–	
Mussel	NA	6.0	18.50	17.25	1.07	0.83	NA	0.06[c]	3.25	–	11.12	13.45	2.04	Soto (1991)
Octopus	NA	6.3	22.00	19.12	1.51	1.25	NA	0.21[c]	10.70	–	0.82	15.82	2.51	Soto (1991)
Fish meal	NA	5.8	55.20	25.80	16.56	16.10	NA	0.22	15.35	7.46	1.70	8.50	0.54	Soto (1991)
Anchovies	–	–	16.98	11.36	4.62	–	0.29	–	–	–	–	92.24	–	Corsino (2016)

[a]Expressed as total nitrogen.
[b]Expressed as total solids (TS) and volatile solids (VS) respectively.
[c]Expressed as phosphate.
NA: not available; TSS: total suspended solids; VSS: volatile suspended solids; tCOD: total chemical oxygen demand; sCOD: soluble chemical oxygen demand.

combination of the different liquid flows constitutes the wastewater produced in the facility and amounts to 14–22 m^3/t of raw fish. Considering that not all of the processing steps always take place simultaneously in the industrial process and that the processed raw material varies throughout time, the characteristics of the produced wastewater are very variable too (Table 6.1).

The largest load is normally present in the effluent from the cookers and the composition depends on the processed product. Fish-canning effluents are characterized by variable composition and flow throughout the operations of each day.

Concerning the wastewater composition, most of the wastewater originating from fish processing activities contains mainly lipids and proteins, except for that originating from mussel processing which is mainly composed of carbohydrates. Organic concentrations range from 10 to 50 g chemical oxygen demand (COD)/L. Temperatures of wastewater flows are variable depending on the fish product and the step of the process where they originate. Heated effluents are produced in the cookers.

With regard to the composition, salinity of the effluents is also highly dependent on the saline content of the processed product and on the fact that seawater is used in the manufacturing plant. In these cases, Cl^-, Na^+, SO_4^{2-}, K^+, Ca^{2+} and Mg^{+2} are the ions present. The concentrations of the majority of the ions are widely variable in the ranges of 8–19 g Cl^-/L, 5–12 g Na^+/L and 0.6–2.7 g SO_4^{2-}/L.

According to the processed product, wastewater may be classified in four categories: mussel cooking wastewater very rich in carbohydrates, specifically glycogen; cephalopods and fish cooking wastewater with a composition depending on the manufacturing system; fish meal production effluents; and wastewater coming from washing, defrosting and cutting, and cleaning with low COD and nutrients concentration.

It is very common for the fish canneries to treat all the effluents together in a wastewater treatment plant, although if possible the separation of those streams with similar compositions helps with the choice of the best applicable treatment (Cristóvão *et al.*, 2012; Carrera, 2020). For those more diluted streams, physical-chemical processes are recommended, however when the stream is highly loaded a combination of these processes with biological ones is required to comply with the required effluent discharge quality.

6.3 CURRENT TREATMENT SYSTEMS

The wastewater originating from fish-canning activities contains large organic matter (mainly lipids, carbohydrates, and sugars), solids, total nitrogen as proteins and salt concentrations. It is characterized by the large concentrations of these pollutants and variable composition due to the different processed raw materials (fish products) that are captured in the fishing activities.

This wastewater type is treated by means of physical-chemical and biological processes (Cristovão *et al.*, 2015a) (Figure 6.1). Usually, for large-size facilities the treatment systems comprise a combination of physical-chemical and biological processes. In this case, the former constitute the primary treatments (equalization, screening, sedimentation, pH adjustment, flocculation, floatation and microfiltration, etc.) while the latter are the core of the soluble organic matter removal treatment and rely on aerobic and anaerobic biological processes. When dealing with small-size industries the physical-chemical processes are preferred due to their low capital costs and ease of operation. Biological processes such as anaerobic digestion and aerobic oxidation are applied for organic matter removal. When nitrogen removal is needed the combined nitrification and denitrification processes are performed in the biological systems (Achour *et al.*, 2000). In these cases, the processes can be implemented either after the anaerobic digester or in an activated sludge system consisting of several units, or a sequencing batch reactor (SBR).

Additionally, the salt content is an important drawback for the treatment and reutilization of the wastewater. For reuse purposes, tertiary treatments are demanded. Among them, membrane systems are applied as they remove salts and pathogens from the wastewater (Cristovão *et al.*, 2015a). Furthermore, ultraviolet (UV) units have been explored in fish canneries to produce pathogen-free effluents.

In the case of large fish canneries, the wastewater treatment system comprises both physical-chemical and biological processes. Initially the fats, oils and greases, and the solids are separated in a dissolved air floatation (DAF) unit combined with coagulation-flocculation, and the obtained effluent is treated in a biological process, either aerobic or anaerobic. The objective of the biological process is to remove the organic matter and, when needed, the nutrients to achieve the discharge limits.

The recovery of some fractions of the wastewater, like the proteins, has also been explored for use in fish flour. For the recovery of fish proteins and oils processes

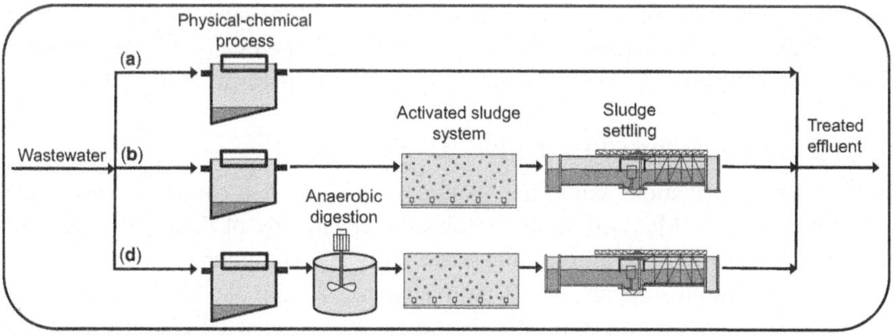

Figure 6.1 Treatment process layouts applied to the treatment of fish-canning effluents.

such as electrolysis, ultrafiltration, reverse osmosis and hydrocyclones, which are relatively expensive processes, are applied.

6.3.1 Physical-chemical processes

Frequently, fish-canning wastewater treatment takes place by physical-chemical processes where the solid fraction and the grease are removed. These processes constitute the wastewater treatment especially in small-size facilities because their construction and operational costs are relatively low. They are installed in large-size factories normally combined with subsequent biological processes. However, nowadays their application is being questioned due to the more stringent normative requirement of more efficient removal treatments, the large surface area they need for implantation, and the huge amounts of solids produced that increase operational costs due to the required management.

Among the different physical-chemical processes that can be applied for the treatment of fish-canning effluents the most commonly employed ones are screening, settling, coagulation-flocculation and dissolved air floatation (Tay *et al.*, 2006). Concerning suspended solids removal, the best results were obtained when a settling step (removing 75% of the fat, oil and grease fraction (FOG) and 48% of TSS) was applied followed by coagulation-flocculation (removing 99.2% of FOG, 85.8% of TSS and 25.2% dissolved organic carbon (DOC)) using $FeCl_3$ (Cristóvão *et al.*, 2014).

Moreover, tertiary treatments necessary to refine the quality of the treated wastewater before discharge are also physical-chemical.

6.3.1.1 Settling

Although in saline conditions, the settling process may experience some drawbacks it is extensively used for the treatment of fish-canning wastewater. The sedimentation of the fish-canning wastewater for relatively short periods (Cristóvão *et al.*, 2014) allows for the separation of three zones: an upper layer comprising the FOG fraction less dense than water; an intermediate clarified zone (approximately 80% of the total volume); and a bottom layer where the solids settle. Removal percentages in the range of 67–75% for FOG and close to 48% for TSS were measured (Aguiar & Sant'Anna, 1988; Cristóvão *et al.*, 2014).

6.3.1.2 Coagulation–flocculation (C–F)

Coagulation-flocculation units are common in the fish-canning wastewater treatment facilities. Many of these facilities, independently of their size, have one of these systems. In general, coagulants are inorganic compounds such as ferrous chloride, aluminium sulphate, etc. (Cristovão *et al.*, 2015a).

Depending on the composition of the treated wastewater the obtained removal percentages are widely variable. Values between 73–99% for FOG and 70–95% for TSS have been reported for coagulant dosages in the range 75–400 mg/L

(Aguiar & Sant'Anna, 1988; Fahim *et al.*, 2001; Cristóvão *et al.*, 2014). Furthermore, the dosage of coagulant and flocculant also varies significantly and as a result so does the mass of separated solids (Cristóvão *et al.*, 2014). In addition, the pH conditions of the coagulation process depend on the type of wastewater as well. In the experiments of Cristóvão (2014) ferric salts were found to be the most suitable coagulants, with dosages of 400 mg/L, even without the need for pH adjustment.

6.3.1.3 Dissolved air floatation (DAF)

Dissolved air floatation systems are meant to remove mainly FOG, which present with low density and tend to float on the water surface. DAF systems base separation of these compounds on the formation of microbubbles formed by pressurized air injected to the recycling water that are released once the stream enters the unit and depressurizes. Thus, the microbubbles help the fats to float. In these systems efficiency percentages found for fish-canning wastewater were 43% of TSS and 94% of FOG (Cristóvão *et al.*, 2014). Important operational parameters are the air supplied pressure and the air/substrate ratio which can vary from 0.058–0.087 kg air/kg FOG for example.

6.3.1.4 Combined processes

It is also common to combine a coagulation-flocculation process with a subsequent DAF. In these cases, the DAF benefits from the fact that the chemical treatment helps to destabilize the repulsion forces among the oil droplets and helps to form larger drops that float more easily. The combination of both processes in a single unit has been explored attaining removal percentages of 41% of TSS and 96% of FOG (Cristóvão *et al.*, 2014).

6.3.1.5 Tertiary treatments

Tertiary treatments are needed for the conditioning of the wastewater before discharge. Recent studies on the feasibility of treated wastewater reuse in the fish canning facilities have been conducted indicating that disinfection is crucial to achieve microbiological quality requirements (Cristóvão *et al.*, 2015a). The use of chlorination for disinfection purposes is widespread, however due to the formation of by-products other treatments based on UV light or reverse osmosis are considered too. (See Chapter 2 for more information on the application of disinfection to saline wastewater.)

6.3.2 Biological processes

Once the solids are removed from the wastewater it is ready to be introduced into a biological treatment system where the soluble organic and nitrogen compounds are eliminated. The applied treatments rely on anaerobic, anoxic, and aerobic processes.

6.3.2.1 Anaerobic digestion

Anaerobic processes are applied for the treatment of fish-canning effluents in particular for effluents with moderate salt content (Table 6.2). They benefit from the wastewater characteristics, the high organic matter concentration, and the temperature (45–60°C). In general, the effluent with larger loads comes from seafood cookers. Effluents produced in fish canneries are highly biodegradable and percentages of 90% biodegradability have been measured in batch experiments (Soto *et al.*, 1991).

In cases where the wastewater presents large salt concentrations, including sulphate, the production of hydrogen sulphide hinders the valorization of the obtained biogas and makes the reactor operation difficult due to the inhibition of the methanogenesis and the associated corrosion events. In this sense, H_2S accumulation in the liquid phase (corresponding to 5% of H_2S in the gas and at pH 7 in the liquid phase) has been found to destabilize the anaerobic digestion (Soto *et al.*, 1991).

Nevertheless, anaerobic digestion has been explored for the treatment of these types of saline wastewater in different kinds of reactors at pilot scale (Balsley-Olesen *et al.*, 1990; Méndez *et al.*, 1992; Omil *et al.*, 1995a). It is important to consider the kind of reactor system applied to each type of wastewater. For example, upflow anaerobic sludge blanket (UASB) reactors have been found to be inappropriate in the case of effluents with large fat concentrations. The COD removal efficiencies are low but can be overcome via the implementation of a pre-hydrolysis step and the control of the alkalinity in the system (Puñal & Lema, 1999). On the other hand, central activity digesters (CAD) arose as a good alternative for the treatment of effluents with concentrations in the range of 15–25 kg COD/m^3 at applied loads of 4 kg COD/($m^3 \cdot d$), attaining COD removal percentages of 76–80% (Méndez *et al.*, 1992). These authors claim that obtained results were possible due to the previous adaptation of the biomass to salt concentrations up to 14 g Cl^-/L. Biofilm systems like upflow anaerobic filters (UAFs) were also tested for the treatment of pre-centrifuged fishmeal wastewater resulting in COD removal percentages of 80% when applied loads were 5 kg COD/($m^3 \cdot d$) (Guerrero *et al.*, 1997). In this case an appropriate recycling ratio of the effluent is needed to overcome inhibitory events caused by large free ammonia concentrations, due to the fact that it improves the mixture facilitating mass transfer rates and minimizing dead zones inside the reactor. Balsley-Olesen (1990) performed experiments with two pilot-scale reactors, an anaerobic filter (AF) filled with shells and an anaerobic fluidized bed (AFB) filled with quartz sand, to treat herring brine with salt concentrations of 10–40 g Cl^-/L. These authors applied an organic loading rate (OLR) of 3–10 kg COD/($m^3 \cdot d$) at salt concentrations in the range of 12 to 21 g Cl^-/L and observed that H_2S in the biogas was 1.5% and the percentage of methane remained constant at 65%. They also claim that the AFB system

Table 6.2 Examples of anaerobic digestion reactors applied for the treatment of fish-canning wastewater operated at mesophilic conditions.

Wastewater Type	Reactor Type	Volume (m^3)	OLR kg COD/ ($m^3 \cdot d$)	Salinity (g Cl^-/L)	COD Removal (%)	Biogas Production $m^3/(m^3 \cdot d)$	Methane (%)	References
Herring brine	AF	0.365	3.3–10	12–21	77–80	1.18–3.09	65	Balsley-Olesen (1990)
	AFB	0.359	3.3–10	12–21	82–86	1.14–5.21	65	
Mussel	AD	0.0009	1.0	13.45	70.1 (69.4[a])	0.027 (0.031)[b]	NA	Soto (1991)
Fishmeal			1.8	8.5	74.4 (43.0[a])	0.042 (0.016)[b]	NA	
Octopus			1.2	15.82	60.0 (25.9[a])	0.033 (0.016)[b]	NA	
Mussel	CAD	15	4.2	14	75–85	8.34	48.1	Méndez (1992)
Tuna	CAD	15	4.5	14	80	5.0	66.6	
Clam	UASB	0.001	1–17	2.8–10.5	80	0.4	70–80	Boardman (1995)
Fishmeal	UAF (PVC rings)	0.00011	5.0	7.5	80	4.0[c]	85–90	Guerrero (1997)
Tuna, Fishmeal, Mussel	CAD	15	4.0	15.0	61.8	0.16	NA	Omil (1995a, b)
Pre-treated fish industry effluents	Mixed-flow AD	0.0015	1.7–0.17	14.6–17.9	70	0.065	50–70	Aspé (1997)

All experiments were carried out at mesophilic conditions 32–37°C.
[a]Thermophilic conditions (55°C).
[b]m^3 $CH_4/(m^3 \cdot d)$.
[c]g CH_4-COD/$(m^3 \cdot d)$.
AF: anaerobic filter; AFB: anaerobic fluidized bed; AD: anaerobic digester; CAD: central activity digester; UASB: upflow anaerobic sludge blanket; UAF: upflow anaerobic filter; PVC: polyvinylchloride; NA: not available.

performs better against shock loads as the mixture conditions inside the reactor are stronger than in the AF, although both handle uneven loads and load interruptions better than suspended-biomass systems.

Omil (1995a) operated a pilot plant comprising a pre-digester for wastewater homogenization, a CAD and a clarifier and managed to treat different mixtures of tuna, fishmeal, and mussel wastewater. A strategy of biomass adaptation to salt and COD load was applied by diluting the influent wastewater, which allowed for the accumulation of average biomass concentrations of 12 g VSS/L capable of dealing with OLR up to 5.5 g COD/($m^3 \cdot$ d) at salinities of 25 g NaCl/L. Approximately 8.2% of the COD fed was removed by the sulphate-reducing bacteria (SRB) activity for H_2S production (Omil *et al.*, 1995b). Methane production from fish processing wastewater, pre-treated in a flocculation unit with ferric chloride, was feasible at a percentage of 50–70% in mixed flow anaerobic digesters operated at a hydraulic retention time (HRT) between 3 and 30 days (Aspé *et al.*, 1997).

Most of the anaerobic treatments take place in a single unit. However, when lipids are present in large concentrations a system comprising two units might be advisable. In these two-stage systems the hydrolysis and acidogenic processes take place in the first unit while acetogenesis and methanogenesis occur in the second one. This process distribution may be accomplished by pH control in acid values or by shortening the HRT or the solid retention time (SRT) lengths. Thus, biomass floatation and inhibition of the methanogenic activity are prevented. In the case of UASB reactors this option aids the granulation process as well. At the same time, it prevents events of pH, temperature, hydraulic or organic shocks.

With respect to operational temperatures, these anaerobic digesters usually operate at mesophilic conditions as at thermophilic ones they are less stable (Soto *et al.*, 1991).

6.3.2.2 Aerobic treatment sludge systems

Aerobic processes are also applied for the treatment of fish-canning wastewater. In general, heterotrophic processes responsible for organic matter removal are resistant to the presence of salts and for this reason these processes perform appropriately in a wide range of operational conditions. Percentages of organic matter removal of 85% have been measured in aerobic systems (Achour *et al.*, 2000) and up to 88% removal has been estimated from the optimization of the process via the application of factorial design (Cristóvão *et al.*, 2015b). Special attention has to be paid to the removal of the FOG fraction as it floats on the surface of the water and significantly affects the oxygen transfer to wastewater, which is crucial for the aerobic biological processes (Chowdhury *et al.*, 2010).

Biofilm systems have been explored for fish processing effluents, as well. As an example, the effluents from marine-products (cuttlefish, octopus and shrimps) processing factories were treated in a fixed bed reactor at laboratory scale at increasing loads from 0.25 to 1.00 kg COD/($m^3 \cdot d$) and with concentrations of 17.6–34.5 g Cl^-/L (Gharsallah *et al.*, 2002). Achieved COD removal percentages were over 56% in all cases.

Additionally, Aloui (2009) succeeded in treating wastewater from a fish industry at an organic loading rate of 0.855 kg COD/($m^3 \cdot d$) in the presence of 40 g NaCl/L using an acclimated sludge.

Nevertheless, the nitrogen removal efficiency in conventional activated sludge (CAS) systems depends on the salt concentration and the degree of acclimation to salt of the sludge used, and it takes place via nitrate if the salt concentration is under 15 g/L (Artiga *et al.*, 2008) but via nitrite for concentrations of around 20 g/L.

The maximum efficiencies obtained in activate sludge systems rely on the operational conditions and wastewater characteristics. Salinity negatively affects the aerobic activity of activated sludge at concentrations over 5–8 g Cl^-/L.

6.3.2.3 Anaerobic ammonium oxidation (anammox) based systems

Nitrogen removal from fish-canning wastewater is usually achieved biologically by the combination of the nitrification-denitrification processes. Recently, the alternative of the combined partial nitritation and anammox processes (PN-AMX) has been explored for this purpose. In this process initially half of the ammonium is oxidized to nitrite and afterwards the remaining ammonium and nitrite are converted to nitrogen gas. Oxygen consumption and organic matter requirements are reduced compared to nitrification and denitrification processes.

There are not many studies on the application of this process to fish processing effluents. Val *et al.* (2018) treated the effluent from an anaerobic digester fed with the effluent generated in a fish cannery and determined that after acclimation of the biomass total nitrogen removal efficiencies of 80% were achieved at nitrogen removal rates of 0.2 kg/($m^3 \cdot d$) at concentrations of 7–9 g NaCl/L. Moreover. The biomass without previous adaptation was able to cope with short-term salt shocks of 16 g NaCl/L for 8 days recovering the initial activity in a few days. Using a similar wastewater type, Gustinianovich (2018) found that the acclimation period of the biomass can be shortened if the amount of used inoculum is large enough to cope with the loss of activity. These authors were able to obtain nitrogen removal efficiencies of 46.3% after 40 days from inoculation with non-adapted to salinity biomass treating a load of 0.029 kg N/($m^3 \cdot d$) at 17.4 g NaCl/L. Additionally they managed to diminish nitrite oxidizing bacteria (NOB) activity by intermittent aeration at salt concentrations below 6 g NaCl/L, which are not inhibitory for this bacterial population.

6.4 INNOVATIVE TECHNOLOGIES

Although in many cases conventional treatments, if operated properly, are suitable for the achievement of treated effluents that comply with the discharge levels fixed by law, frequently factories do not have enough space to implement them. In these cases, the development of innovative technologies characterized by their low implantation surface requirements is an objective. Among the possible systems, those based on biofilm biomass or membranes provide this advantage. In addition, stricter discharge levels are proposed with the aim of protecting the receiving water bodies and this implies the need for implementation of more efficient nutrient removal systems.

The following innovative technologies have been tested with this type of effluent to cope with the existing drawbacks.

6.4.1 Aerobic granular sludge systems

Aerobic granular sludge (AGS) systems have been explored as an alternative technology to treat effluent from the fish-canning industry (Figueroa *et al.*, 2008; Val del Río *et al.*, 2013; Corsino *et al.*, 2016; Carrera *et al.*, 2019). In most of these studies, aerobic granular biomass was produced in sequencing batch reactors (SBRs) operated in cycles comprising a short feeding period, an aerated reaction period, a settling period, and an effluent withdrawal period (Table 6.3). Scarce information is available on the application of AGS systems operated with anaerobic plug feeding and aerobic phases (Carrera *et al.*, 2019). The operational conditions imposed in the SBR aim at enhancing the biomass aggregation in the form of granules based on the establishment of a feast/famine regime. In fully aerobic AGS systems (Af) organic matter and nitrogen are removed simultaneously, while in the anaerobic fed AGS (Anf) reactors phosphorus can be also removed together with COD and N.

The comparison of the performance of two AGS systems fed with fish-canning wastewater, under aerobic and anaerobic conditions, respectively, indicated that the granular stability is better in the former case (Carrera *et al.*, 2019). Furthermore, nitrogen removal is higher in this case while phosphorus removal only takes place in the reactor with anaerobic feeding.

AGS systems can be operated for the purpose of nitrogen removal. Due to salt concentrations of the wastewater, nitrogen removal takes place via nitrite as NOB activity is more sensitive than ammonium oxidizing bacteria (AOB) activity to these conditions. AOB activity was observed to take place, in experiments using synthetic media, at 33 g NaCl/L (Bassin *et al.*, 2011) and 50 g NaCl/L (Wan *et al.*, 2014) while NOB activity was completely inhibited at 20 g NaCl/L (Pronk *et al.*, 2014). Nitrite accumulation was observed even at low salt concentrations in systems fed with nitrogen-rich wastewater (Figueroa *et al.*, 2008). In this study, a combined effect of free ammonia concentration (up to 3 mg N/L) and salinity (26.5–28.3 g NaCl/L) was proposed as the responsible factor for the accumulation of nitrite observed in the reactor.

In general, the removal of organic matter and nitrogen is feasible in AGS reactors. The formation of granular biomass is possible, and its characteristics depend on the

Table 6.3 Summary of aerobic granular (AGS) reactors treating fish-canning wastewater.

Wastewater Type and Composition (mg/L)	Reactor Type (Feeding Mode)	Volume (m^3)	AGS Properties	Applied Loading Rate $kg/(m^3 \cdot d)$	Removal Efficiency (%)	References
Diluted effluent from an anaerobic digester treating fish-canning wastewater tCOD: 410 sCOD: 350 NH_4^+-N: 45 NaCl: 6,600	SBR (Af)	0.0015	Size: 3.4 mm Density: 60 g VSS/L_{gran} TSS: 9.3 g/L SVI: 27.9 mL/g TSS	COD: 1.46 NH_4^+-N: 0.18	COD: 90–95 N: 20–45	Figueroa (2008)
Effluent from a seafood industry pre-treated in a DAF tCOD: 870 sCOD: 785 NH_4^+-N: 90 TSS: 63 Conductivity 2.9 mS/cm	SBR (Af)	0.0018	Size: 3.4 11 mm Density: 18–35 g VSS/L_{gran} TSS: 7.0 g/L SVI: 45–91 mL/g TSS	COD: 4.4 NH_4^+-N: 0.45	COD: 90 N: 30	Val del Río (2013)
Anchovy cannery tCOD: 16,984 sCOD: 11,362 TN:1,152 NaCl: 50,000	SBAR (Anf)	0.0035	Size: 2.1 mm Density: 25.6 g TSS/L_{gran} TSS: 4 g/L	COD: 5.4	COD: 90 N: 95 PM: 80	Corsino (2016)
Fish-canning wastewater after the DAF sCOD: 610–1,710 NH_4^+-N: 72–106	SBR (Af)	0.00175	Size: 1.1–2.52 TSS: 2–12 g/L SVI_5: 27–70 mL/g TSS	COD: 1.8–6.6 N: 0.22–0.45	COD: 70–87 N: 32–90	Carrera (2019)
TSS: NaCl: 5,000–13,300	SBR (Anf)	0.00175	Size: 0.89–1.63 TSS: 2.0–7.5 g/L SVI_5: 68–130 mL/g TSS		COD: 70–93 N: 17–40	

PM: particulate matter; Anf: anaerobic feeding; Af: aerobic feeding; SBAR: sequencing batch airlift reactor; SVI: sludge volume index; SVI_5: sludge volume index measured after 5 minutes of settling.

substrate composition. The presence of coagulants coming from the pre-treatment produces less dense granules which hinder the stability of the granular bed (Val del Río *et al.*, 2013). While the efficiency of COD removal is high, close to 90%, the nitrogen removal relies on the salt concentration and eventually on the free ammonia concentration achieved depending on the ammonia concentration and pH of operation (Figueroa *et al.*, 2008).

The best feeding strategy, Af or Anf, was evaluated in two different reactors fed with the same batch of fish-canning wastewater with the result that the formation of granular biomass lasted longer in the Af reactor than in the Anf one (Carrera *et al.*, 2019). Although, in the Af system the granules obtained better resisted the changes in OLR and salt concentrations. Moreover, even if the COD removal was similar, in the case of nitrogen removal it took place via nitrification-denitrification in the Af reactor and only due to assimilation in the Anf reactor. In the latter, P removal was not possible by phosphorus-accumulating organisms (PAO) due to the salt concentrations of the feeding.

6.4.2 Membrane bioreactors

Only a few studies provide results on the application of membrane systems for the biological treatment of or product recovery from fish-canning wastewater (Table 6.4).

Afonso and Bórquez (2002) evaluated the recovery of proteins from the effluents generated in a fishmeal factory, with conductivities of 3.1–5.2 μS/cm, using an ultrafiltration (UF) membrane. Despite the low protein rejection of approximately 3–39%, the obtained results were useful and allow identification of possible improvements such as the use of a UF membrane with smaller pore size or their substitution by nanofiltration (NF) membranes.

With regard to their use for wastewater treatment, Artiga (2005) tested an innovative membrane-assisted hybrid bioreactor for the treatment of fish-canning wastewater. This system comprised a hybrid circulating bed reactor (CBR), where the biomass grew on plastic particles, coupled in series to an ultrafiltration membrane module (UFM). In this system organic matter and nitrogen were removed (Table 6.4). Nitrification (50–60% of the reactor capacity) took place mainly in the biofilm biomass while heterotrophs were mainly present in suspension. The fact that a settler is not needed for biomass retention, due to the presence of the membrane results in the requirement of a smaller surface for implantation compared to CAS systems. Membrane bioreactors (MBRs) appear as appropriate systems for the treatment of high saline wastewater as they can deal with saline concentrations up to 83 g/L (Artiga *et al.*, 2008) providing efficiencies of 95% COD removal treating loads of 1.4 kg COD/($m^3 \cdot d$). However, nitrogen removal is not feasible at these saline conditions.

Sridang (2006) applied a MBR for the treatment of seafood-processing wastewater and observed that the response of the system treating wastewater with

Table 6.4 Examples of aerobic membrane reactors applied for the treatment of fish-canning effluents.

Wastewater Type	Reactor Type	Volume (m^3)	OLR kg $COD/m^3 \cdot d$	NLR kg NH_4^+-$N/m^3 \cdot d$	Salinity (g NaCl/L)	COD Removal (%)	References
Mixture of influent and effluent from an industrial anaerobic contact reactor	CBR-UFM	0.0055–0.0012	6.5	1.8	NA	99%	Artiga (2005)
Seafood processing wastewater	MBR	0.035	0.28	0.016	NA	97	Sridang (2006)
Effluent from tuna cooking in brine tanks	Pilot Hybrid MBR	0.5 (0.2 anoxic, 0.2 aerobic, 0.1 filtration chambers)	0.3–1.4	N.A.	27–89[a]	65–92	Sridang (2006)
Effluent from tuna cooking with steam injection			0.1–4.0	0.08–0.70[b]	20[a]	50–90%	

CBR: circulating bed reactor; UFM: ultrafiltration module; NA: not available; NLR: nitrogen loading rate.
[a]As conductivity (mS/cm).
[b]Nitrogen expressed as TN.

or without salt was similar, with COD removals up to 97% while total Kjeldahl nitrogen (TKN) removal was limited to 5.3%. Nonetheless, these authors propose the implementation of subsequent treatments, oxidation, and adsorption, to decrease the COD concentration of the permeate for further reuse and recycling purposes.

A specific drawback of the use of membrane systems for the treatment of saline wastewater is the appearance of inorganic scaling of compounds like calcium phosphate or undissolved NaCl (Artiga *et al.*, 2008). This necessitates frequent cleaning cycles for the membrane maintenance.

6.5 CASE STUDY: WASTEWATER TREATMENT PLANT IN A FISH CANNING FACILITY

6.5.1 Fish-canning wastewater

Most of the canned seafood in the world is tuna. Spain is one of the major producers of canned tuna, approximately two-thirds in terms of volume, which accounts for almost 64% of the production in the EU (EUMOFA, 2019). Furthermore, the biggest factories are in the Autonomous Community of Galicia (NW of Spain). Due to the importance of the sector, one of these facilities was chosen as a model to develop the present case study.

The selected fish canning facility processes, as raw, 5,000 tons of fresh tuna per year. The wastewater is produced during fish thawing, washing, and eviscerating, fish and cans cooling and washing, cooking, and clean-up of washing areas. Considering the process efficiency of 55%, 2750 tons of canned tuna is produced per year. The volume of wastewater produced is 8 m^3 per ton of tuna canned (Henze *et al.*, 1997), thus 22,000 m^3/y of wastewater is produced. The flowrate of wastewater generated is 88 m^3/d, taking into account that the facility operates only five days a week, eight hours a day (Uttamangkaborvorn *et al.*, 2005), and the corresponding free days each year.

Before discharge, this produced wastewater needs to be treated in the treatment plant designed for this purpose.

The discharge limits of the different pollutants present in the produced effluent depend on the discharge point (Table 6.5). Less stringent limits for salts are set in those cases where the sewage collector is the destiny of the effluent. Stringent limits exist only when freshwater is the receiving body or the effluent is destined for reuse. When it is discharged to the sea, no limitation for salt concentrations is defined.

The administration, normally the council in which the factory is located, is responsible for fixing the discharge limits of the different parameters contained in the produced effluent. For this reason, there are no fixed limits for seawater discharge, but they are normally in the range indicated in Table 6.5. In addition,

Table 6.5 Example of discharge limits according to the discharge point of effluents generated in the treatment plants of fish canning plants (from: IHOBE, 2019).

Parameter	Unit	To Sewage Collector[1]	To the Sea[2]
pH	–	6–10	5.5–9.5
Temperature	°C	40	$rT < 3$[3]
BOD_5	mg/L	–	50–200
COD	mg/L	1,500	100–400
TSS	mg/L	500	100
Chloride	mg/L	2,000	–
Ammonium nitrogen	mg/L	50	20–80
Nitrite nitrogen	mg/L	20	–
TKN	mg/L	50	–
Total Phosphorus	mg/L	50	5–20
Sulphate	mg/L	500	–
Oils and fats (floating)	mg/L	100	10–50
Sulphide	mg/L	2	0.5–2.0

[1]Recommended value, in each case the fish cannery will check on the limit values established by the wastewater treatment plant (WWTP) operator. [2]Recommended values not included in any legislation. [3]The increase of temperature will not exceed 3°C at 100 m from the discharge point. BOD_5: biological oxygen demand measured after 5 days.

Augas de Galicia is the body in charge of the surveillance of the fulfilment of these limits in the Autonomous Community.

In the present case, the release to seawater in the middle of an estuary (named ría in Galicia) by means of a discharge pipe was selected (Table 6.5). This is the most common option in this autonomous community due to the factories being located on the coast and the fact that seawater is used in the process and therefore saline effluents are produced. In this case, no limits for chloride and sulphate concentrations are provided.

6.5.2 Applied wastewater treatment processes

A medium-size facility producing canned tuna was selected for this case study. This factory produces 2750 tons of canned tuna per year. The flowrate of wastewater produced is fixed at 88 m^3/d. Seawater is used as processing water in some of the steps of the process, like cleaning or cooking, and supplies salinity to the wastewater. Due to the activities in the plant two types of wastewater are produced characterized by their load (Table 6.6).

The high load stream accounts for 5% in volume of the total wastewater and contains most of the load. It comes from the tuna cookers and it is mixed with

Table 6.6 Representative composition of high and low load wastewater samples collected in a tuna canning facility which uses seawater as process water employed for the case studies (using data from Carrera (2020) as reference).

Parameter		High Load	Low Load
Flowrate	m^3/d	4.4	83.6
Conductivity	mS/cm	18.0	20.9
pH	–	6.59	5.91
Total suspended solids	mg/L	2,370	225
Volatile suspended solids	mg/L	1,070	217
Total COD	mg/L	10,800	1,183
Soluble COD	mg/L	7,300	691.2
Greases	mg/L	1,600	311.41
Volatile fatty acids	mg COD/L	3,300	44.46
Ammonia	mg N/L	560	3.4
Total nitrogen	mg N/L	1,020	18.73
Phosphate	mg N/L	130	2.1
Proteins	mg/L	2,060	102.76
Carbohydrates	mg/L	510	72.31
Chloride	mg Cl^-/L	4,210	8,410
Sulphate	mg SO_4^{-2}/L	250	1,130
Sodium	mg Na^+/L	2,770	4,490

other liquid streams generated in the process containing fats and oils. It contains the highest concentration of solids, organic matter, and nitrogen. The low load stream corresponds to 95% in volume of the total wastewater produced and it is less concentrated than the previous stream, except for the fact that it contains large salt concentrations. It is produced mainly in cleaning activities.

Four scenarios have been assessed: (A) both streams are treated together via physical-chemical processes; (B) both streams are treated separately combining physical-chemical and aerobic biological processes; (C) both streams are treated separately combining physical-chemical, aerobic and anaerobic biological processes; and (D) like case C but including nitrogen removal.

The four proposed treatment schemes must fulfil the following restrictions:

- Preferably, the wastewater types will be treated separately to avoid the generation of large loads containing relatively high pollutant concentrations from the mixture of high and low loaded wastewater types.
- The effluent produced from a unit must fulfil the requirements of the next unit in the treatment line. As an example, coarse solids should be removed before the DAF or biological treatment systems, or the COD concentration should be

large enough for an anaerobic treatment, or an intense FOG removal is necessary before the biological processes.

- All the schemes needed the implementation of a sieving system to remove coarse solids from low and high load wastewater types.
- The size of the resulting systems should not be so large as to represent a drawback for the plant.

The treatment units comprising the train processes of each scenario are described below.

Scenario A: For the low load wastewater the system comprises a sieving system (particles larger than 1 mm), a dissolved air floatation (DAF) unit and a coagulation-flocculation unit.

Scenario B: Both streams are treated separately in two treatment lines. The low load wastewater is treated in three subsequent units: sieving, DAF and coagulation-flocculation units. The high load wastewater flows through a sieving unit, a DAF unit, and an activated sludge system for organic matter removal. The effluents from both treatment lines are mixed before being discharged.

Scenario C: Both streams are treated separately in two treatment lines. The low load wastewater is treated in three subsequent units: sieving, DAF and coagulation-flocculation units. The high load wastewater is treated in a sieving and a DAF unit for solid and fat removal and then it is anaerobically digested. The effluents from both treatment lines are mixed before being discharged.

Scenario D: The same as in Scenario C but including nitrogen removal in an activated sludge (AS) system.

Four scenarios have been proposed based on existing industrial facilities or experimental set-ups from previous research studies.

The removal percentages corresponding to each treatment have been chosen as the highest possible values assuming the appropriate performance of each unit (Table 6.7).

Table 6.7 Removal percentages selected for the different wastewater treatment processes applied in the four scenarios evaluated (from: IHOBE, 2019).

	BOD_5 (%)	TSS (%)	Fats (%)
Sieving	15	25	30
DAF	30	28	80
Settling	45	70	50
Coagulation-flocculation	70	90	50
Activated sludge	90	90	0
Activated sludge with nitrogen removal	90	90	0
Anaerobic digestion	80	90	90

6.5.3 Alternative evaluation

6.5.3.1 Technological evaluation

Calculations were performed for the four cases, with the aim to remove the pollutants present in each wastewater type (as summarized in Table 6.6). These are detailed in Annex 1 (Tables A1.1 to A1.4). Obtained results are presented in Table 6.8 where the discharge limits are provided.

Assumptions made for calculations (Table 6.8):

- Volatile suspended solids concentration for the discharge limit conditions was estimated as 80% of the TSS limit concentration.
- Soluble COD concentration for the discharge limit conditions was estimated from the concentration of total COD minus the concentration of COD corresponding to the VSS using the ratio of 1.5 g COD/g VSS.
- In the solid separation units, the ratio VSS/TSS is the same for the solids separated and remaining in the treated wastewater flow.
- The removal of soluble COD occurred in the same percentage as the BOD_5 considering that most of it corresponds to biodegradable COD.
- The percentage of total COD removal was estimated the same as for the TSS and VSS but considering only the non-soluble fraction.
- The percentage of nitrogen removal in the solid removal processes was not considered as in all the scenarios its concentration was below the discharge limit.
- According to Balsley-Olesen (1990) 1.5% of H_2S can be present in the biogas, which will correspond to a concentration of 0.054 mg H_2S/L in the effluent from the AD according to Henry's law. This concentration was estimated using the Henry's law constant of 525.5 ± 16.4 atm/mol fraction (Yongsiri *et al.*, 2005). This concentration is only relevant in Scenario C where this treatment is the last one before high load wastewater discharge. In Scenario D the H_2S is degraded in the subsequent activated sludge system.

All scenarios evaluated produced effluents in the range of concentrations required for discharge. In the case where nitrogen limits are more restricted the implementation of an activated sludge system with nitrogen removal for the treatment of high load wastewater will reduce its concentration by 80% in the final effluent. The pH value of all the streams was within the ranges defined by the discharge limits. Only in the options with coagulation-flocculation could its value be significantly different from the desired value.

Cristóvão (2014) proposed that the best system for solid removal from fish-canning wastewater is a sedimentation followed by a coagulation-flocculation stage; while if the FOG fraction is to be removed they do not recommend the use of DAF treatments for these effluents even if they are combined with coagulation-flocculation. Nevertheless, in Galicia many fish canneries have a DAF system forming part of the physical-chemical wastewater

Table 6.8 Effluent quality of the four options compared to the chosen discharge limits for the case of using a discharge pipe in the sea.

Parameter	Units	Discharge limit	Scenario A	Scenario B	Scenario C	Scenario D
TSS	mg/L	100	17.9	17.9	17.9	12.2
VSS	mg/L	80	14.0	14.0	14.0	11.4
Total COD	mg/L	100–400	108.2	108.2	108.2	76.1
Soluble COD	mg/L	280	182.4	138.9	160.6	121.6
Grease	mg/L	10–50	26.3	31.9	21.8	21.8
Ammonia	mg N/L	20–80	31.2	31.2	31.2	6.0
Total nitrogen	mg N/L	–	68.8	68.8	68.8	22.9
P total	mg P/L	5–20	8.5[1]	8.5[1]	8.5[1]	8.5[1]
Chloride	mg Cl^-/L		8,200	8,200	8,200	8,200
Sulphate	mg SO_4^{-2}/L		1,086	1,086	1,086	1,086
Sulphide	mg H_2S/L	0.5–2.0	–	–	0.0027	–
Sodium	mg Na^+/L		4,404	4,404	4,404	4,404

In Scenario A it corresponds to the global effluent from the two combined streams treated. In Scenarios B, C and D, it represents the composition of the combination of both streams after their separate treatment and considering the respective flowrates.

[1]Phosphorus as phosphate.

treatment. Thus, no unique options are applicable for the treatment of fish-canning wastewater and each case needs to be studied individually.

Although Scenario D is the most complex one, it presents advantages such as the fact that the presence of an anaerobic digester before the aerobic unit diminishes odour emission (Achour *et al.*, 2000).

6.5.3.2 Cost evaluation

A rough calculation of the operational costs of the four scenarios assessed in terms of applied technology was performed based on the type of treatment and the wastewater flow. Energy consumption of the different treatments were taken from Metcalf and Eddy (2014) referred to the volume of treated wastewater (Table A1.5 in Annex 1). Some assumptions were made for calculations (Table 6.9) and are as follows:

- The fish cannery operates 251 days per year.
- The price of the kWh was fixed at 0.11 € (Eurostat, 2019).
- The reagent chosen for coagulation-flocculation was ferric chloride with dosages of 400 mg/L and the corresponding cost of 290 €/t according to Cristóvão (2014)
- The level of sludge recycling in the secondary treatment was 0.3 m^3 recycled/m^3 treated.
- The biomass productivity in the AS system was estimated as 0.35 g of VSS produced/g COD removed. In addition, the ratio 0.75 g VSS/g TSS was considered for the solid production calculation.
- The solid separated from the physical-chemical processes and the sludge purged from the AS were fed to a gravity thickener where the obtained concentration was 45 g TSS/L (Roibás-Rozas *et al.*, 2020).
- The valorization of the produced methane is not included as the quality of the generated biogas is too low due to its H_2S content and would require the implementation of an expensive post-treatment (Barros *et al.*, 2009).

From the results the cost of the coagulant in the four cases used to treat the mixed or low loaded flows is significant. It represents approximately 35% of the total estimated operational costs which makes it almost futile to try to optimize the other treatments to enhance the quality of the discharged wastewater. Furthermore, even if this cost was not considered here, the mass of solid waste which must be managed increases in proportion to the added coagulant and contributes to the operational costs of the treatment plant as well by 56–63%.

Thus, looking from another point of view it should not be a problem to implement more stringent treatments if the coagulation process is improved as there is enough room in the huge costs of this treatment to compensate for the new implemented treatments. This is good news in terms of environment protection.

Table 6.9 Results from the operational cost calculations corresponding to the four scenarios evaluated.

	Units	SCENARIO A	SCENARIO B		SCENARIO C		SCENARIO D	
Flow Type		**Mixed**	**High Load**	**Low Load**	**High Load**	**Low Load**	**High Load**	**Low Load**
Wastewater flow	(m^3/d)	88	4.4	83.6	4.4	83.6	4.4	83.6
Wastewater influent pumping	kWh/d	3.39	0.17	3.22	0.17	3.22	0.17	3.22
Screens	kWh/d	0.04	0.00	0.03	0.00	0.03	0.00	0.03
Activated sludge for BOD removal	kWh/d	–	0.62	–	–	–	–	–
Activated sludge with nitrification/denitrification	kWh/d	–	–	–	–	–	1.01	–
Return sludge pumping	kWh/d	–	0.02	–	–	–	0.07	–
Secondary settling	kWh/d	–	0.02	–	–	–	0.02	–
Dissolved air floatation	kWh/d	3.08	0.15	2.93	0.15	2.93	0.15	2.93
Mesophilic anaerobic digestion (primary plus waste activated sludge)	kWh/d	–	–	–	0.56	–	0.56	–
Total energy consumption	kWh/d	6.50	0.98	6.18	0.88	6.18	1.98	6.18
OPERATIONAL COSTS (energy)	€/year	180	27	171	24	171	55	171
OPERATIONAL COSTS TOTAL (energy)	€/year	180	198		195		226	
Reagent consumption for coagulation-flocculation ($FeCl_3$)	t/d	0.035	–	0.033	–	0.033	–	0.033
Solids from physical-chemical processes	t/d	0.063	–	0.056	–	0.056	–	0.056
Biological solids from AS	t/d	–	0.008	–	–	–	0.002	–
OPERATIONAL COSTS (solid management)	€/year	4,733	4,824		4,219		4,340	
OPERATIONAL COSTS TOTAL	€/year	7,475	7,455		6,848		6,999	

6.6 CONCLUSIONS

Among the technologies applied for the treatment of fish-canning wastewater those based on physical-chemical processes for solid and FOG fraction separation and biological processes for soluble COD removal are the most widely used.

Solids removal is needed to fulfil the discharge requirements and to obtain a stream appropriate to be treated in the biological processes performed either in aerobic or anaerobic conditions.

The most important compound to be removed in the physical-chemical processes is the grease as it is not easily removed in the biological processes.

Nitrogen removal is not required to fulfil the current discharge limits, however in the case where the regulations became more stringent to limits near those from urban WWTP technologies, its removal should be considered in the treatment facilities.

Salt is not removed from these effluents. In most cases when the effluent contains salt due to the employment of seawater in the production process it is because the plant is located near to the coast. In these cases, the treated effluent is discharged back to the sea.

With respect to the operational costs, those associated with the coagulant acquisition are the most important ones. The operation of the treatment technologies represents only 2.4–3.2% distributed as 0.4–0.8% for the high load wastewater and 2.3–2.5% for the low loaded flow in cases B–D. Improving the solid separation steps is crucial to decrease these costs in the fish canneries.

REFERENCES

Achour M, Khelifi O., Bouazizi I. and Hamdi M. (2000). Design of an integrated bioprocess for the treatment of tuna processing liquid effluents. *Process Biochemistry*, **35**, 1013–1017.

Afonso M. D. and Bórquez R. (2002). Review of the treatment of seafood processing wastewaters and recovery of proteins therein by membrane separation processes-prospects of the ultrafiltration of wastewaters from the fish meal industry. *Desalination*, **142**, 29–45.

Aguiar A. L. C. and Sant'Anna G. L. (1988). Liquid effluents of the fish canning industries of Rio de Janeiro state – treatment alternatives. *Environmental Technology Letters*, **9**, 421–428.

Aloui F., Khoufi S., Loukil S. and Sayadi S. (2009). Performances of an activated sludge process for the treatment of fish processing saline wastewater. *Desalination*, **246**, 389–296.

Artiga P., Oyanedel V., Garrido J. M. and Méndez R. (2005). An innovative biofilm-suspended biomass hybrid membrane bioreactor for wastewater treatment. *Desalination*, **179**, 171–179.

Artiga P., García-Toriello G., Méndez R. and Garrido J. M. (2008). Use of a hybrid membrane bioreactor for the treatment of saline wastewater from a fish canning factory. *Desalination*, **221**, 518–525.

Aspé E., Martí M. C. and Roeckel M. (1997). Anaerobic treatment of fishery wastewater using a marine sediment inoculum. *Water Research*, **31**(9), 2147–2160.

Balsley-Olesen P., Lynggaard-Jensen A. and Nickelsen C. (1990). Pilot-scale experiments on anaerobic treatment of wastewater from a fish processing plant. *Water Science and Technology*, **22**(1/2), 463–474.

Barros M. C., Magán A., Valiño S., Bello P. M., Casares J. J. and Blanco J. M. (2009). Identification of best available techniques in the seafood industry: a case study. *Journal of Cleaner Production*, **17**, 391–399.

Bassin J. P., Pronk M., Muyzer G., Kleerebezem R., Dezotti M. and van Loosdrecht M. C. M. (2011). Effect of elevated salt concentrations on the aerobic granular sludge process: linking microbial activity with microbial community structure. *Applied Environmental Microbiology*, **77**(22), 7942–7953.

Boardman G. D., Tisinger J. L. and Gallagher D. L. (1995). Treatment of clam processing wastewaters by means of upflow anaerobic sludge blanket technology. *Water Research*, **29**(6), 1483–1490.

Carrera P. (2020). Aerobic Granular Sludge Technology for Fish-canning Wastewater Treatment: Optimisation and Scale up. PhD Thesis, USC.

Carrera P., Campo R., Méndez R., Di Bella G., Campos J. L., Mosquera-Corral A. and Val del Rio A. (2019). Does the feeding strategy enhance the aerobic granular sludge stability treating saline effluents? *Chemosphere*, **226**, 865–873.

Chowdhury P., Viraraghavan T. and Srinivasan A. (2010). Biological treatment processes for fish processing wastewater – a review. *Bioresource Technology*, **101**(2), 439–449.

Corsino S. F., Capodici M., Morici C., Torregrosa M. and Viviani G. (2016). Simultaneous nitritation-denitritation for the treatment of high-strength nitrogen in hypersaline wastewater by aerobic granular sludge. *Water Research*, **88**, 329–336.

Cristóvão R. O., Botelho C. M. S., Martins R. J. E. and Boaventura R. A. R. (2012). Chemical and biological treatment of fish canning wastewaters. *International Journal of Bioscience, Biochemistry and Bioinformatics*, **2**(4), 237–242.

Cristóvão R. O., Botelho C. M., Martins R. J. E., Loureiro J. M. and Boaventura R. A. R. (2014). Primary treatment optimization of a fish canning wastewater from a Portuguese plant. *Water Resources and Industry*, **6**, 51–63.

Cristóvão R. O., Botelho C. M., Martins R. J. E., Loureiro J. M. and Boaventura R. A. R. (2015a). Fish caning industry wastewater treatment for water reuse – a case study. *Journal of Cleaner Production*, **87**, 603–612.

Cristóvão R. O., Gonçalves C., Botelho C. M., Martins R. J. E., Loureiro J. M. and Boaventura R. A. R. (2015b). Fish canning wastewater treatment by activated sludge: application of factorial design optimization. Biological treatment by activated sludge of fish canning wastewater. *Water Resources Industry*, **10**, 29–38.

Eurostat. (2019). See https://ec.europa.eu/eurostat/statistics-explained/index.php?title=File:Electricity_prices_for_non-household_consumers,_first_half_2020_(EUR_per_kWh)_v4.png (accessed 24 February 2021).

EUMOFA. (2019). Monthly Highlights, No. 6. See https://www.eumofa.eu/documents/20178/151768/MH+6+2019+EN+final.pdf (accessed 24 February 2021).

Fahim F., Fleita D., Ibrahim A. and El-Dars F. S. (2001). Evaluation of some methods for fish canning wastewater treatment. *Water Air Soil Pollution*, **127**, 205–226.

Figueroa M., Mosquera-Corral A., Campos J. L. and Méndez R. (2008). Treatment of saline wastewater in SBR aerobic granular reactors. *Water Science and Technology*, **58**(2), 479–485.

Ghaly A. E., Ramakrishnan V. V., Brooks M. S., Budge S. M. and Dave D. (2013). Fish processing wastes as a potential source of proteins, amino acids and oils: a critical review. *Journal of Microbial and Biochemical Technology*, **5**, 107–129.

Gharsallah N., Khannous L., Souissi N. and Nasri M. (2002). Biological treatment of saline wastewaters from marine-products processing factories by a fixed-bed reactor. *Journal of Chemical Technology and Biotechnology*, **77**, 865–870.

Giustinianovich E. A., Campos J.-L., Roeckel M. D., Estrada A. J., Mosquera-Corral A. and Val del Río A. (2018). Influence of biomass acclimation on the performance of a partial nitritation-anammox reactor treating industrial saline effluents. *Chemosphere*, **194**, 131–138.

Guerrero L., Omil F., Mthdez R. and Lema J. M. (1997) Treatment of saline wastewaters from fish meal factories in an anaerobic filter under extreme ammoniacal concentrations. *Bioresource Technology*, **61**, 69–78.

Henze M., Harremoës P., Jansen J. C. and Arvin E. (1997). Wastewater Treatment. Biological and Chemical Processes, 2nd ed. Environmental Engineering series. Ed. Springer-Verlag, Berlin. ISBN: 978-3-662-22607-0

IHOBE S. A. (2019). White Book for the waste and emissions minimization. Fish canneries (In Spanish: Libro blanco para la minimización de residuos y emisiones. Conserveras de pescado). https://www.azti.es/wp-content/uploads/2019/05/Libro-blanco-para-la-minimizaci%C3%B3n-de-residuos-en-conserveras_azti.pdf (accessed 24 February 2021).

Mendez R., Omil F., Soto M. and Lema J. M. (1992). Pilot plant studies on the anaerobic treatment of different wastewater from a fish-canning factory. *Water Science and Technology*, **25**(1), 37–44.

Metcalf and Eddy. (2014). Wastewater Engineering: Treatment and Resource Recovery, 5th edn. McGraw-Hill, New York.

Omil F., Méndez R. J. and Lema J. M. (1995a). Characterization of biomass from a pilot plant digester treating saline wastewater. *Journal of Chemical Technology and Biotechnology*, **63**, 384–392.

Omil F., Méndez R. and Lema J. M. (1995b). Anaerobic treatment of saline wastewaters under high sulphide and ammonia content. *Bioresource Technology*, **54**, 269–278.

Pronk M., Bassin J. P., De Kreuk M. K., Kleerebezem R. and Van Loosdrecht M. C. M. (2014). Evaluating the main and side effects of high salinity on aerobic granular sludge. *Applied Microbiology and Biotechnology*, **98**(3), 1339–1348.

Puñal A. and Lema J. M. (1999). Anaerobic treatment of wastewater from a fish-canning factory in a full-scale upflow anaerobic sludge blanket (UASB). *Water Science and Technology*, **40**(8), 57–62.

Roibás-Rozas A., Mosquera-Corral A. and Hospido A. (2020). Environmental assessment of complex wastewater valorisation by polyhydroxyalkanoates production. *Science of the Total Environment*, **744**, 140893.

Soto M., Méndez R. and Lema J. M. (1991). Biodegradability and toxicity in the anaerobic treatment of fish canning wastewaters. *Environmental Technology*, **12**, 669–677.

Sridang P. C., Kaiman J., Pottier A. and Wisniewski C. (2006). Benefits of MBR in seafood wastewater treatment and water reuse: study case in southern part of Thailand. *Desalination*, **200**, 712–714.

Suvanich V., Ghaedian R., Chanamai R., Decker E. A. E. A. and McClements D. J. (2006). Prediction of proximate fish composition from ultrasonic properties: catfish, cod, flounder, mackerel and salmon. *Journal of Food Science*, **63**, 966–968.

Tay J.-H., Show K.-Y. and Hung Y.-T. (2006). Wastewater Treatment in the Food Processing Industry. Taylor & Francis, Boca Raton, Fla, USA.

Uttamangkabovorn M., Prasertsan P. and Kittikun A. H. (2005). Water conservation in canned tuna (pet food) plant in Thailand. *Journal of Cleaner Production*, **13**, 547–555.

Val del Río A., Figueroa M., Mosquera-Corral A., Campos J. L. and Méndez R. (2013). Stability of aerobic granular biomass treating the effluent from a seafood industry. *International Journal of Environmental Research*, **7**(2), 265–276.

Val del Río A., Pichel A., Fernández-González N., Pedrouso A., Fra-Vázquez A., Morales N., Mendez R., Campos J. L. and Mosquera-Corral A. (2018). Performance and microbial features of the partial nitritation-anammox process treating fish canning wastewater with variable salt concentration. *Journal of Environmental Management*, **208**, 112–121.

Veiga M. C., Méndez R. and Lema J. M. (1994). Anaerobic filter and DSFF reactors in anaerobic treatment of tuna processing wastewater. *Water Science and Technology*, **30**(12), 425–432.

Wan C., Yang X., Lee D. J., Liu X., Sun S. and Chen C. (2014). Partial nitrification of wastewaters with high NaCl concentrations by aerobic granules in continuous flow reactor. *Bioresource Technology*, **152**, 1–6.

Xiao Y. and Roberts D. J. (2010). A review of anaerobic treatment of saline wastewater. *Environmental Technology*, **31**(8–9), 1025–1043.

Yongsiri C., Vollertsen J. and Hvitved-Jacobsen T. (2005). Influence of wastewater constituents on hydrogen sulfide emission in sewer networks. *Journal of Environmental Engineering*, **131**(12), 1676–1683.

Chapter 7

Case study: Urban wastewater treatment plant with saline intrusions

ABSTRACT

In this chapter the performance of a wastewater treatment plant (WWTP) treating wastewater with saline content due to seawater intrusions is evaluated. Aspects regarding the effects on the removal efficiency and performance of the different processes are considered as well as the solutions put into practice for the different drawbacks associated with the salt presence. In particular, the increase of reagents needed in the physical-chemical processes, the production of H_2S due to anaerobic biological processes and the decrease of biomass activity will be evaluated. Two case studies are assessed based on the salinity of the wastewater treated (3,000 and 7,000 μS/cm) in a WWTP installed on the Mediterranean coast of Spain. For both case studies, two possible scenarios are analysed: the direct discharge of the effluent to the sea or its reuse for agriculture irrigation.

Keywords: Biomass activity, hydrogen sulphide, precipitation, solid production

7.1 MOTIVATION

Water scarcity is one of the challenges to be faced in the future in order to ensure the availability of freshwater for the population of the world. Of course, the reduction of water consumption should be considered as a first action. Independently of the water availability this resource has to be considered as limited and consumed wisely. In particular, because after its use a proportional amount of wastewater is produced that requires further treatment. A second action is to recover the water fraction

Author(s): Anuska Mosquera Corral, Ángeles Val del Río and José Luis Campos Gómez

doi: 10.2166/9781789060645_0151

from the produced wastewater by regenerating it in wastewater treatment plants (WWTPs). For this purpose, appropriate treatments need to be applied depending on the quality required for reuse.

Among the characteristics of the regenerated water influencing its further reuse the salinity needs to be considered. An important amount of the produced wastewater in the world contains salt concentrations which correspond to electrical conductivity (EC) values over 3,000 μS/cm (total dissolved solids (TDS) = 1.92 g/L) in the range of moderately saline water (Rhoades *et al.*, 1992). The volume of saline wastewater is expected to grow mainly in coastal areas due to the sea level increase caused by the global warming effect or the exploitation of seawater in substitution of freshwater. When saline wastewater is regenerated the conventional treatments applied in the existing WWTP are not able to decrease the salt concentration as, generally, these facilities were not designed with consideration of this aspect. In some cases, although the regenerated water is not of high quality it can be used for irrigation with some limitations. This is the case for effluents with EC values below 3,000 μS/cm. It has been estimated that 6% of the world cultivated land has saline conditions generally associated with the presence of NaCl (Pedrero *et al.*, 2018). In other cases salts need to be removed or recovered in posterior treatment stages for further reuse. Thus, when the aim is to regenerate saline wastewater the treatment plants should be designed to obtain the effluent quality required for each application. In the case where good quality regenerated wastewater, in terms of salt concentration, is produced it will help to guarantee the availability of enough freshwater for the future world population reducing the increasing pressure on natural resources.

7.2 OCCURRENCE OF SALINE URBAN WASTEWATER

Urban wastewater is expected to be produced mainly from the use of freshwater in domestic activities, which already provide certain saline conditions due to decalcifiers and cleaning additives. However, there are cases in which saline streams are mixed with this freshwater generating wastewater with saline characteristics. As a consequence the regenerated water obtained after wastewater treatment usually contains significant salt concentrations (chlorides, sulphates, etc.).

Among the streams, sources of salinity are the seawater infiltrated in the sewage systems of WWTPs located in coastal areas (Flood & Cahoon, 2011), the discharge of saline industrial wastewater to collector systems (Guerreiro *et al.*, 2020) or, in more limited cases, the use of seawater to flush toilets in substitution of freshwater (Tang *et al.*, 2006). On average seawater intrusions have an EC of 8,000 μS/cm, but this value will depend on the volume of seawater intruded. As a result of its mixture with urban sewage, the produced wastewater presents an EC over 5,000 μS/cm. However, this kind of wastewater does not represent, in general, the largest percentage of all the produced wastewater.

As an example, in the Mediterranean region of Murcia Spain, located on the Mediterranean coast, from the 100 Hm^3 of treated wastewater used for irrigation per year, 93% has an EC over 2,000 μS/cm and 37% has EC values over 3,000 μS/cm (Pedrero *et al.*, 2012). The highest fraction of 56% presents values between of EC between 2,000 and 3,000 μS/cm and although the quality is not too high it can be reused for irrigation purposes. Only 7% of the produced wastewater has an EC below 2,000 μS/cm. For those fractions of wastewater with saline contents over those admitted for irrigation, the application of desalination processes is required.

Wastewater containing up to 5,000 μS/cm is produced from other types of intrusions that contribute to supplying salt to the wastewater, such as in residential areas located on the coast where saline groundwater levels are high and water leaks in basements have to be bailed out. This water flows into the collecting system. Furthermore, salt is present already in the wastewater due to the characteristics of the freshwater employed and the products used for washing activities. In addition, some industries discharge their saline effluents to the sewage system as well.

In coastal areas the amounts of produced saline wastewater are increasing in volume and nowadays it is estimated that 9% of the total produced wastewater is saline.

In order to understand the magnitude of the saline effluents in the world, the amount of the population living in coastal areas must be considered. In this sense, according to the SEDAC (Socioeconomic Data and Applications Center) approximately 40% of the population in the world lives within 100 km of the coast. Furthermore, an estimation indicates that 80% of the largest WWTPs are located in coastal areas. Considering the large population of these areas, the food requirements and the fact that generally agriculture is the most important industrial sector in the region, the amount of water resources becomes an important issue.

For these reasons, in recent decades the reuse of water from WWTPs has increased significantly (Hochstrat *et al.*, 2005; Levine & Asano, 2004). Probably the most frequent reuse is in agriculture due to several advantages of these streams which contain nutrients, reduce fertilizer application and improve the management of the effluents in a more effective way (Candela *et al.*, 2007).

7.2.1 Facing seawater intrusions

In practice most of the wastewater entering sewage systems comprises a variable fraction of waterflow due to infiltrations associated with groundwater levels. An estimated amount of infiltration referred to the area served by the collection system may range from 0.2 to 28 $m^3/ha \cdot d$ (Metcalf & Eddy, 2003). The infiltration flow relies on several factors: (i) the quality of the material, workmanship and building connections of the collection system; (ii) the maintenance routines; and (iii) the level of the groundwater in comparison with that of the collection system. The rate and amount of infiltration varies according to the population densities, which will define the length of the collection system and the area served, and also to the type of soil and the topographic conditions. Besides infiltrations, those sewage systems with leakage problems also present exfiltration events. These ones are beyond the scope of this chapter and only infiltration will be considered.

In addition to the previous considerations, WWTPs that have been designed to treat non-saline wastewater end up needing to cope with seawater intrusions and consequently operational problems arise. Variable sea levels associated with variable-intensity tides are responsible for these saline intrusions in collection systems. Therefore, several responses are observed in the WWTPs such as diminishment of the treatment efficiency, increase of water flows resulting in bypass establishment when exceeding the plant capacity, and occurrence of mechanical effects resulting in increasing operational and maintenance costs (Flood & Cahoon, 2011). Among them, those associated with damage due to corrosion and loss of activity in the biological systems result in important reductions of removal efficiency. Furthermore, the fact that salts are present, as in the case of sulphate, enhances the development of new processes not considered in the facility design such as the occurrence of sulphate reduction to sulphide.

Saline infiltrations occur to a major or minor extent in all countries with coastal areas.

7.2.2 Seawater used for toilet flushing

In those areas in the world where water is scarce several alternative options to solve this limitation are applied. Among them the use of seawater with minimal treatment (pre-filtration and chlorination) for toilet flushing is an interesting option. In Hong Kong this has been a normal practice since the 1950s (Li *et al.*, 2005) and it is in place for 80% of its population (Leung *et al.*, 2012). Its utilization is being explored for implementation in other zones, in particular those which are located 30 km away from the seashore and are highly populated (Liu *et al.*, 2016), based on a triple water supply (TWS) that comprises fresh water supply, seawater for toilet flushing, and grey water reuse and recycling. Additionally, the specifically developed process of Sulphate reduction, Autotrophic denitrification and Nitrification Integrated (SANI) is being studied as well with the aim to reduce overall costs by 50%, energy consumption by 35% and greenhouse gas (GHG) emissions by 36% (Chen *et al.*, 2012). As a consequence saline wastewater needs to be treated. Several studies have proven the adequacy of these practices apart from the economic and environmental benefits obtained (Tang *et al.*, 2006). The application of this alternative requires changes in the urban management systems and relies on the existence of a dual water supply system where seawater is only used to flush toilets while freshwater is used only for potable uses.

In order to determine the economic advantage of the seawater use, a study performed by Tang *et al.* (2006) evaluated the engineering costs of this alternative in comparison with the utilization of raw freshwater or reclaimed water. These authors evaluated six scenarios of water supply for toilet flushing. Among them three consisted of the use of separate seawater, freshwater and reclaimed water, respectively. Obtained results indicated that using seawater for toilet flushing provided the best results (reduction of 44% and 49% of the freshwater and reclaimed water costs, respectively) compared to the use of fresh or reclaimed water.

Although this seems a very advantageous alternative it is not widespread. For this reason in the next section only the specific case of WWTPs that receive sewage comprising seawater due to marine intrusions in the collection systems is considered.

7.2.3 Saline industrial wastewater discharge

The salinity associated with industrial saline wastewater discharge depends on the location and the industrial activity of the zone. For this reason, its contribution is highly variable and a representative composition for this contribution is not available.

Among the different sources of industrial saline wastewater which are dumped to wastewater collector systems are those from cheese production (120,000 μS/cm), detergent production (56,700 μS/cm), the meat industry (35,600 μS/cm), fruit canneries (25,700 μS/cm), etc. These effluents need to comply with the concentration limits set for discharge into collector systems.

7.3 TREATMENT CONSIDERATIONS IN WWTPS

Depending on the saline conditions of the urban wastewater entering the WWTP the effects on the different treatment processes will be different as well as the treatment solutions to be applied (Table 7.1). In general, salinity affects the activity of the biological processes (in particular nitrification), solids separation processes (as it affects settleability, floatation, dehydration), disinfection and odour production (due to H_2S production). Furthermore, the fact that saline intrusions enter the collection system results in an increase in the volume of wastewater to be treated in the WWTP.

The main aspects related to the influence of salinity on these processes are summarized in Table 7.1. The solution action adopted or proposed is indicated as well. A more extensive analysis of this information is provided in the text below where the most relevant effects of the salt presence are analysed.

Table 7.1 Summary of the effects of salinity on the different treatment process and full-scale adopted solutions (see chapters 2, 3 and 4 for a more detailed description of the different processes).

Process	Effect of salinity	Effect on performance	Solution
Sewage collector	• Odour production associated with H_2S formation. • Corrosion occurrence. • Human toxicity.	• COD valorization for methane production diminishes. • GHG emission increases.	• Reduce the time spent in the collectors by increasing the slope of the collection pipes. • Improve mixing conditions and/or aeration in aerobic bioreactors. • Modify the environment for biological processes to occur.
Settling	• Sludge floatation due to the increase in density of the water associated with the salt concentration.	• Sludge volume index decreases. • Turbidity of the supernatant increases.	• Removal of floating solids from the surface of the settler.
Floatation	• Enhances the formation of small air bubbles.	• Particle-air bubble collisions.	• Reduce the chemical coagulants addition.
Coagulation-flocculation	• Affects the performance of the added coagulants.	• Necessity for higher or lower dosage of reagents.	• Adjust the added dosage of reagents.
Secondary treatment	• Decrease of biomass activity, in particular, the nitrification process.	• Biological reactor efficiency decreases thus the effluent quality is not complied with.	• Adjust the HRT. • Adjust the SRT from 10 to 15 days in order to favour the nitrification process.
Secondary treatment	• Biomass floatation due to H_2S production.	• Solid retention inside the reactor decreases and consequently the efficiency of the process diminishes. • Development of filamentous bacteria.	• Implement a biomembrane reactor in substitution of the conventional activated sludge system.

(*Continued*)

Table 7.1 Summary of the effects of salinity on the different treatment process and full-scale adopted solutions (see chapters 2, 3 and 4 for a more detailed description of the different processes) (*Continued*).

Process	Effect of salinity	Effect on performance	Solution
Secondary settling	• Floatation of biomass associated with N_2 gas formation.	• Sludge volume index decreases. • Turbidity increases.	• Control the retention time in the settler to avoid the production of H_2S from sulphate reduction with COD from the biomass in anaerobic conditions. • Shorten retention time in the settler to avoid denitrification. • Substitute settling with a membrane system.
Anaerobic digestion	• Decrease of methanogenic activity. • Production of H_2S at inhibitory levels for the biomass. • Presence of H_2S in the biogas.	• Decrease of anaerobic digester efficiency.	• Adjust the SRT from 10 to 15 days in order to favour sludge anaerobic digestion.
Effluent quality	• Salinity exceeds the concentration suitable for reuse in agriculture irrigation.	• Impossibility of reclaimed water reuse.	• Implement a tertiary treatment based on RO membranes.
Membrane filtration	• Precipitation of salts as their solubility is exceeded in the proximity of the membrane.		• More intense washing of the membrane.
Membrane filtration	• Biofouling caused by biomass particle decrease due to salinity.	• Increase of energy consumption for membrane operation. • Less duration of membranes.	• Acclimate biomass to saline wastewater to avoid negative effects of salinity

COD: chemical oxygen demand; GHG: greenhouse gases; HRT: hydraulic retention time; SRT: sludge retention time; RO: reverse osmosis.

7.3.1 Hydrogen sulphide production

Wastewater reaches the WWTP through a collection system which receives the effluents produced in a certain region with a surface extension which depends on the kind of urban nuclei (residential, with small industry, touristic, etc.), the population density and the size of the WWTP. Thus, the time of permanence of the wastewater in the collection system relies on the length of the system and constructive characteristics. The collection systems comprise among others force mains, gravity sewers, manholes, pumping equipment, etc.

In general hydrogen sulphide (H_2S) is naturally formed in wastewater systems from the anaerobic degradation of organic matter (Eq. 7.1). Anaerobic conditions occur frequently in the collection systems and other units of the plant where mixture and aeration conditions are inefficient. This compound represents a major drawback in the wastewater network (Lens *et al.*, 1998) and is responsible for odour and corrosion (due to H_2SO_4 formation) production and health impacts on humans (Churchill & Elmer, 1999; Jiang *et al.*, 2015, 2016; Li *et al.*, 2017; Nielsen *et al.*, 1992). Apart from the release of sulphur from proteins in saline wastewater, sulphates are also present in large concentrations and contribute to hydrogen sulphide production.

$$C_{10}H_{19}O_3N + 6.25\ SO_4^{2-} + 9.375\ H^+ \rightarrow 3.125\ H_2S + 3.125\ HS^- + 7\ H_2O + 9\ CO_2 + NH_4^+ + HCO_3^- \quad (7.1)$$

In these conditions if the wastewater residence time inside the collectors is long enough anaerobic conditions occur which promote the reduction of sulphates to H_2S. Besides in the collector, H_2S is produced in other zones in the WWTP where anaerobic conditions and presence of sulphates and organic matter coexist.

Low wastewater velocities inside the collection systems correspond to long residence times and favour the formation of microbial biofilms on their surfaces which are responsible for H_2S production (Carrera *et al.*, 2016). To avoid the occurrence of this compound several actions have been proposed which rely either on collection pipe construction or modification of the existing environment to prevent H_2S production or remove it after its formation. With respect to the former one, in the design of the collection system it is crucial to provide pronounced slopes that enhance the water flow reducing wastewater permanence time inside the collectors. When dealing with systems of new construction this alternative is feasible but it is complicated to achieve with existing ones.

With respect to the environment modification, the existing alternatives are also applicable to other units in the plant. This case requires the addition of an external compound into the collector system or WWTP unit, like nitrate, ferrous salts, injection/addition of air, pure oxygen, or alkali, to increase pH or metals to precipitate sulphide (Kulandaivelu *et al.*, 2020). As an example, Churchill and Elmer (1999) tested the addition of nitrate which competes for the organic matter with sulphate reducing bacteria (SRB) avoiding H_2S formation. In this case, nitrate was supplied in a pilot test into a collector section, fitted with H_2S meters in the Town of Bedford (Massachusetts, USC). The authors observed that shortly after its addition the measured concentration of H_2S was zero and concluded that the configuration of the collection systems will define the number of dosing stations needed to achieve this situation.

The reaction that takes place in the collection system with nitrate addition, and using the wastewater as organic matter source, is (Eq. 7.2):

$$0.02\,C_{10}H_{19}O_3N + 0.1175\,SO_4^{2-} + 0.0021\,NO_3^- + 0.0021\,H^+ \rightarrow 0.0021\,C_5H_7O_2N + 0.0588\,H_2S + 0.0588\,HS^- + 0.1336\,H_2O + 0.1693\,CO_2 + 0.0200\,NH_4^+ + 0.0200\,HCO_3^- \quad (7.2)$$

All the possible options have drawbacks associated with energy and maintenance requirements for air injection, reagents consumption due to the need for dosage in multiple locations or increase of nitrate load, flocculation and settling of solids (in sewer pipes) after iron salt dosing, etc. The iron dosing was found to be one of the most applied options from a survey performed in Australia (Ganigué *et al.*, 2018) as it is not very expensive and does not produce harmful by-products.

Moreover, the dosage of iron either in the collection system or the inlet of the WWTP provided beneficial effects in other stages of the process. Its dosage in the collection network provided iron sufficient for desulfurization of H_2S via iron sulphide precipitation from biogas in the anaerobic sludge digester (Ge *et al.*, 2013; Salehin *et al.*, 2019). Kulandaivelu *et al.* (2020) evaluated its dosage in the sewer pipes and in the inlet of the of the Luggage Point WWTP (Queensland, Australia) and observed the decrease of H_2S concentration in the gas phase in the anaerobic digester (36–45%) and an improvement in the digested sludge dewaterability of almost 10%.

The production of H_2S during solid anaerobic digestion of the produced sludge in the WWTP is of great importance. Information about this process has been provided in Chapter 4.

7.3.2 Solid separation

High salt concentrations favour the action of air bubbles on the floatation of particles and as a consequence decrease the requirements for chemical coagulant addition (Quinn *et al.*, 2007). However, the presence of salt reduces the quality of the effluent after settling due to the appearance of turbidity although the effect on the SVI (sludge volume index) is to decrease it, indicating an improvement of the settling properties of the biomass. Chen *et al.* (2018) observed a decrease in SVI from 95 to 19.5 mL/g TSS (total suspended solids) when the salinity increased from 0 to 20 g NaCl/L. Further explanation on the reasons for this behaviour is provided in Chapter 2 of this book.

In the case of the flocculation process, increasing salt concentrations up to certain values improved the process, but over these values it was negatively affected as observed by Wang *et al.* (2013) when they applied it to separate TiO_2 nanoparticles at concentrations over 5.85 g NaCl/L and 3.33 g $CaCl_2$/L. When the formation of flocs is based on overcoming the solubility of the formed hydroxide the presence of salts exerts a negative effect. Wang *et al.* (2010) observed that the efficiency decreased when $AlCl_3$, PAC (polyaluminium chloride) and nano-PAC were used to separate humic acids and the concentration of salt augmented from 2.925 to 14.625 g NaCl/L. On the contrary when coagulation occurs due to polymers addition the coagulation is improved by salt addition. In the case of humic substances (HS) (10–40 mg/L) their removal improved with the decrease of salinity from 25 to 10 g NaCl//L and was 89 and 73%, respectively (Hilal *et al.*, 2008). With respect to the coagulation with aluminium sulphate, salinity decreases the concentration of the dosage (0.862–6.840 mg $Al_2(SO_4)_3$/L) required for organic substances coagulation (Duan *et al.*, 2002).

Membrane filtration in the WWTP is a common practice to obtain an effluent from the secondary treatment with enough quality for reuse in agriculture. The use of ultrafiltration (UF) membranes (0.04 μm pore size) for the treatment of secondary effluents from WWTPs enabled the removal of faecal pathogenic organisms, without the need for a disinfection step, and the production of reclaimed wastewater appropriate for agricultural reuse to irrigate tomato, fennel and lettuce (Lopez *et al.*, 2010). These authors observed that the microbial pollution in the reclaimed water was in some cases below that of the freshwater due to the existence of illegal discharge of untreated wastewater to aquifers. For concentrations of salt below 10 g NaCl/L conventional membrane bioreactor (MBR) systems are applicable (Tan *et al.*, 2019). Salinity affects the sludge activity and characteristics, by diminishing the biomass particle size leading to an increase of the filtration resistance in membrane systems associated with the development of membrane fouling (Xie *et al.*, 2014). Thus salinity makes the operation of these reactors more difficult and more research is needed to find out the exact process taking place.

7.3.3 Biological processes

The effects of salinity on the secondary treatment performance are the decrease of biological processes activity, cell death and biomass floatation with consequent wash-out. The bacterial populations involved in the different removal processes are affected differently. There is not a general consensus regarding the resistance of activated sludge to salinity. Some authors fixed the limit of salt concentration at 1% (Abou-Elela *et al.*, 2010) while for others activated sludge was able to adapt to salt concentrations up to 5% (Lefebvre and Moletta, 2006). In addition, rapid shifts in salt concentrations, as in the case of seawater intrusions, diminish the efficiency of the biological processes in WWTPs (Salvadó *et al.*, 2001) unless the biomass is acclimated to saline conditions, then, the negative effect comes when the salt concentration is decreased (He *et al.*, 2017). For concentrations under 10 g NaCl/L microorganisms are capable of acclimating while for concentrations above 30 g NaCl/L the time required is too long or acclimation is not possible (Linarić *et al.*, 2013). If the sludge comprises non-halophilic microorganisms these might need an acclimation period to perform properly (Tan *et al.*, 2019). However, in the case of transient saline conditions this acclimation might not possible (Flood & Cahoon, 2011). In order to overcome the decrease of the biomass activity the operation SRT is increased. As a consequence, the pumping of returning sludge increases with increasing reactor volume. Considering that the higher the SRT the lower the observed biomass yield, the net amount of biomass generated in these conditions diminishes. This is a benefit in terms of cost of solid management.

Nitrifying bacteria are affected by saline conditions to different extents depending on the time of exposure to saline conditions. Ammonium oxidizing bacteria (AOB) are more sensitive to short and long term exposure to 10 g Cl^-/L than nitrite oxidizing bacteria (NOB). According to Moussa *et al.* (2006) their activity diminished by 36 and 39% and 5 and 13% for short and long term exposure, respectively.

Salinity also exerts other negative effects like the increase of effluent turbidity (Salvadó *et al.*, 2001), after the secondary settler, caused by the decrease of diversity of eukaryotic microorganisms (protozoa and small metazoa) (Salvadó *et al.*, 2001), which normally graze on the particulate material and bacterial cells. In the case of nitrifiers, the increase of salt resulted in an improvement of the biomass settleability and reduction of the proportion of filamentous bacteria (Moussa *et al.*, 2006).

From the previous analysis it is clear that the effect of salinity can be either positive, inexistent or negative with respect to the correct performance of the studied processes (He *et al.*, 2017). However, the addition of salts is not considered in general to improve any of these processes as these compounds are not beneficial for further reuse, although they are not a problem in the case of treated effluent discharged to the sea.

7.4 CASE STUDY: URBAN WASTEWATER TREATMENT PLANT

7.4.1 Case studies definition

The WWTP example chosen for this evaluation was designed with a treatment capacity of 130,000 person equivalent and a treated flow of 20,000 m^3/d (0.231 m^3/s). Thus, a person equivalent was estimated to produce 154 L/d of wastewater to calculate the wastewater flowrate.

The facility is located on the Mediterranean coast of Spain and as a consequence receives transient intrusions of seawater due to its infiltration in the collection system. Data from the WWTP of San Pedro de Pinatar, Autonomous Community of Murcia, were used as examples for the definition of the case study (Cadagua & Ferrovial-Agromán, 2007). Currently the discharge point of the plant is on the coast through a discharge pipe and presents a limited surface for implantation that imposes the need for the implementation of compact treatment systems. In this study the reuse of regenerated wastewater for agriculture irrigation purposes was evaluated as an alternative to discharge, in those cases where reclaimed water of sufficient quality was produced.

The WWTP chosen comprised the following treatment steps (Figure 7.1):

- Primary treatments: catchpit, pumping, coarse screeners, degritting and degreasing and fibre-removal strainers.
- Secondary treatment: activated sludge system with extended aeration supplied with a membrane bioreactor for ultrafiltration of the effluent and sludge separation that returns to the inlet of the biological reactor in a ratio 4:1.
- Sludge treatment: the sludge produced in excess is thickened and dehydrated and sent to an authorized management company.
- Tertiary treatment: ultraviolet disinfection unit.
- Effluent desalination: reverse osmosis treatment (only in Case B).

The salt content of the wastewater entering the plant for the proposed case studies was chosen at two different values depending on the supposed seawater infiltration

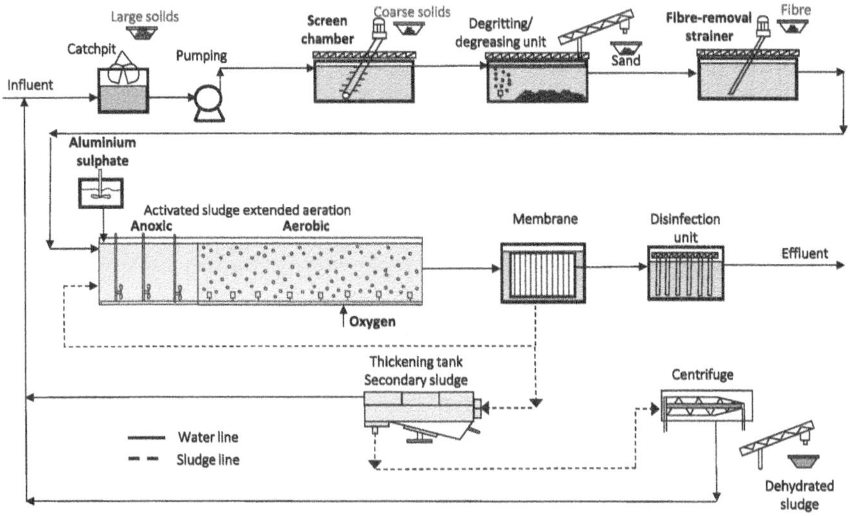

Figure 7.1 Layout of the wastewater treatment plant example.

ratio: 3,000 fixed μS/cm, which is the maximum value allowed to be present in water suitable for agriculture irrigation, and 7,000 μS/cm, which means that produced regenerated water is not suitable for irrigation without desalination treatment.

The composition of the wastewater is detailed in Table 7.2. Some calculations and assumptions were made to define all the parameters of the wastewater necessary for each case study and the plant discharge limits, and these are indicated below and in the table footer.

The concentration of salts in the wastewater was estimated from the TDS concentration minus the concentration of soluble organic matter (sCOD). The sCOD was calculated as the total chemical oxygen demand (tCOD) minus the COD associated with the volatile suspended solids (VSS) of the wastewater as 130 mg sCOD/L, using the ratio 1.8 kg COD/kg VSS of primary settled solids. The concentration of sCOD as TDS was calculated with the stoichiometric expression that defined the COD composition of the wastewater, estimated as follows: (i) 400 mg tCOD/L were converted to carbon using the ratio 12 mg C/32 mg O (150 mg C/L), (ii) thus, the stoichiometric ratio of C, N and P was determined from the concentration 150 mg C/L, 40 mg N/L and 7 mg P/L as 5, 1.15 and 0.082 mol, respectively, (iii) the stoichiometric parameters of H and O were defined, taking as a basis the expression C_5H_9NO, as 9 and 1, respectively (Mateju *et al.*, 1992). From these calculations, the stoichiometry $C_5H_9N_{1.14}P_{0.09}O$ ($Mw = 103.8$ g/mol) was estimated as defining the sCOD concentration of the wastewater. Applying this procedure the soluble organic matter concentration as solids was calculated as 224.9 mg DS/L.

Afterwards, the concentration of salts was estimated from the remaining concentration of TDS and applying the ratio between the ions existing in the seawater composition, defined according to the standard ASTM D 1141-98 (Table A2.1 in Annex 2). Considering that salinity from origins other than seawater intrusions contributed in low proportion, only the ratio among the ions in the seawater was used to determine their values for the wastewater and the discharge limits (Table 7.2).

The defined discharge limit concentrations for water reuse were fixed according to the Directive 91/271/CEE and the data provided by the Food and Agriculture Organization of the United Nations (FAO) as appropriated for agriculture irrigation (FAO, 1985). The optimum value of EC for irrigation was found to be below 2,000 μS/cm which was chosen to define the discharge limit for irrigation reuse (Zarzo *et al.*, 2013).

The two different case studies evaluated and compared were defined in terms of the EC of the wastewater and the installed treatment processes as follows:

- Case A: treated wastewater (Table 7.2) with a conductivity of 3,000 μS/cm. The treatment plant design included primary treatments, a secondary one comprising an extended aeration bioreactor with a membrane ultrafiltration

Table 7.2 Composition of the wastewater (Cadagua & Ferrovial-Agromán, 2007) and discharge limits fixed by the directive 91/271/CEE and the FAO limits for irrigation (FAO, 1985).

Parameter	Unit	Wastewater Composition Case A	Wastewater Composition Case B	Discharge Limits for Water Reuse
pH	–	–	–	6.5–8.4$^{(4)}$
Electrical conductivity	μS/cm	3,000	7,000	2,000
TDS	mg/L	1,920$^{(1)}$	4,480$^{(1)}$	1,280$^{(1)}$
BOD_5	mg/L	180	180	25
tCOD	mg/L	400	400	125
TSS	mg/L	200	200	35
Ammonium	mg/L			
Nitrite	mg N/L			
Total N	mg N/L	40	40	10
Total P	mg P/L	7$^{(2)}$	7$^{(2)}$	1
Chloride	mg Cl^-/L	940$^{(3)}$	2,360$^{(3)}$	1,065$^{(4)}$
Boron	mg B^-/L	0.224$^{(3)}$	0.56$^{(3)}$	2$^{(4)}$
Sodium	mg Na^+/L	522$^{(3)}$	1,311$^{(3)}$	920$^{(4)}$
Sulphate	mg SO_4^{2-}/L	131$^{(3)}$	329$^{(3)}$	960$^{(4)}$
Magnesium	mg Mg^{2+}/L	63$^{(3)}$	158$^{(3)}$	61$^{(4)}$
Calcium	mg Ca^{2+}/L	20$^{(3)}$	50$^{(3)}$	400$^{(4)}$
Potassium	mg K^+/L	19$^{(3)}$	47$^{(3)}$	2$^{(4)}$
Oils and fats (floating)	mg/L	90$^{(2)}$	90$^{(2)}$	20
Sulphide	mg/L			1
Total coliform	No./100 mL	10^7–10^9 $^{(2)}$	10^7–10^9 $^{(2)}$	
Faecal coliform	No./100 mL	10^4–10^6 $^{(2)}$	10^4–10^6 $^{(2)}$	
Cryptosporidium oocysts	No./100 mL	10^{-1}–10^1 $^{(2)}$	10^{-1}–10^1 $^{(2)}$	$\leq 1^{(5)}$
Giardia lamblia cysts	No./100 mL	10^{-1}–10^2 $^{(2)}$	10^{-1}–10^2 $^{(2)}$	

The salinity of the discharge limit was fixed at 2,000 μS/cm as a value of salinity suitable for reuse of the regenerated water in agriculture.
[1]Estimated from the ratio TDS (mg/L) = EC (μS/cm) · 0.64 (Metcalf & Eddy, 2003).
[2]Data from Metcalf & Eddy (2003) for medium strength wastewater type.
[3]Estimated from the composition of artificial seawater model (ASTM D 1141-98) after subtracting 259.5 mg/L of organic matter contained in the wastewater.
[4]Limits from the FAO (1985).
[5]Intestinal nematodes (helminth eggs): ≤1 egg/L for irrigation of pastures or forage.
BOD_5: 5-day biological oxygen demand.

system, and a tertiary disinfection treatment. The produced sludge was concentrated and dried for further management.

- Case B: treated wastewater (Table 7.2) with a conductivity of 7,000 μS/cm. The wastewater treatment plant comprised the same treatments as in Case A, and in addition the produced effluent was submitted to a tertiary treatment which includes salt removal via a reverse osmosis (RO) process.

In both cases two options were evaluated, the discharge of the obtained effluent to the sea (A1 and B1, respectively) and the valorization of the reclaimed water for irrigation (A2 and B2, respectively). This option of reuse is a reasonable application as in the Mediterranean coastal area, where the plant is installed, approximately 70% of the WWTP effluents are valorized for this purpose.

The quality of the effluent will be maintained the same in all case studies, in terms of nutrients concentration (nitrogen and phosphorus), in order to comply with the discharge limits imposed by the Directive 91/271/CEE in the case of sensitive areas (Table 7.2). Although in the case of agriculture application these nutrient concentrations could be higher to exploit the fertilization potential of the wastewater, this option was not considered here. This choice was made due to the special situation of eutrophication of the Mar Menor (Murcia) where the irrigation water is supposed to run-off from the production fields receiving the reclaimed water.

The case studies were compared in terms of energy consumption and operational costs. Energy consumption associated with aeration, pumping and membrane systems operation was considered. Operational costs were estimated considering energy consumption and solids management. Produced sludge was not valorized for methane production as the plant used as an example did not have this treatment. Other operational costs such as personnel and maintenance were not considered as no significant differences were expected for this cost in the different scenarios evaluated. These costs contribute to approximately 30% and 10% of the total costs, respectively.

After defining the case studies, the following assumptions were made for the calculations referring to the performance of the activated sludge system. The equations used are described in Annex 2:

- For calculations of the net waste activated sludge produced (Annex 2, Eq. A2.1), the total COD contained in the wastewater (400 mg tCOD/L) was fractionated using the following ratios: 3 kg S_S/kg S_i and 3 kg X_S/kg X_i (estimated from the composition of the wastewater in different countries from Henze *et al.* (2002)), 1.8 kg COD/ kg VSS (primary settled solids), 1.42 kg COD/kg VSS (secondary settled sludge). The obtained concentrations were: 140.25, 159.75, 46.75 and 53.25 mg COD/L for Ss (readily biodegradable soluble COD fraction), Xs (slowly biodegradable particulate COD fraction), Si (non-biodegradable soluble COD fraction) and Xi (non-biodegradable particulate COD fraction), respectively.

- Previous calculated biodegradable particulate and soluble COD fractions were removed in the extended aeration biological reactor in order to comply with the Directive 91/271/CEE limits (Table 7.2). In practice for calculations the concentration of tCOD in the effluent was below this limit due to the need for an electron donor to denitrify the nitrogen content of the wastewater and the presence of the UF membrane. Thus, the effluent quality was fixed equal to the composition corresponding to the example WWTP of 3 mg SS/L, 17 mg tCOD/L and 2 mg BOD_5/L.
- In the pre-treatment steps 60% of the TSS of the wastewater was removed with a content of 26.6% inerts (Eq. A2.2 and A2.3). This 60% was then added to the TSS produced in the plant and that required posterior management. For biomass growth only 40% of this fraction was degraded by the heterotrophic bacteria (Eq. A2.1).
- The organic matter from the influent wastewater was removed in the activated sludge reactor via denitrification (6.6 kg COD/kg NO_3^--N reduced, according to Metcalf & Eddy, 2003) and the amount left was oxidized in the aeration zones.
- The total load of daily produced sludge waste from the plant was estimated as the produced kg VSS/d from the COD biodegraded minus the kg VSS/d removed in the effluent plus the solids separated in the pre-treatments (Eq. A2.3). The flow of VSS was then converted into TSS by assuming that the fraction of VSS corresponded to 75% of the fraction of TSS. The price of the solid management was fixed as 200 €/t TSS.
- All nitrogen present in the water was hydrolysed to ammonium, which was nitrified and subsequently denitrified to achieve a concentration of 10 mg TN/L in the effluent.
- The oxygen consumed was the sum of that required for the oxidation of ammonium to nitrate (4.33 kg O_2/kg N oxidized (Werzernak & Gannon, 1967)) and that needed for the oxidation of the excess organic matter for denitrification (Eq. A2.4).
- The price of the reclaimed water for reuse was 0.3 €/m^3.
- Calculations were performed with the SRT fixed for winter and summer periods at 25 and 18 days, respectively.
- The energy consumed for pumping the wastewater to the activated sludge system was estimated from the inflow (Case A) and from the inflow plus the amount of seawater infiltrated (Case B). In both cases the energy consumption was 0.045 kWh/m^3. In the case of the sludge recycled from the MBR this value was 0.013 kWh/m^3. The recycling ratio (R) was fixed at 4.
- Reverse osmosis was chosen to remove the salinity in case study B when the regenerated wastewater was treated for reuse in irrigation.
- The energy consumption attributed to the ultrafiltration system was 0.3 kWh/m^3 of permeate and to the reverse osmosis without energy recovery was 0.65 kWh/m^3 of permeate (Metcalf & Eddy, 2014).

- The price of the kWh was fixed at 0.11 € (Eurostat, 2019).
- In Case B with an EC value of 7,000 μS/cm a correction percentage due to salinity, calculated to counteract its effect on oxygen solubility (S_{O2}) and oxygen mass transfer coefficient (K_{La}), was applied to estimate the oxygen required for the secondary treatment (Annex 2). The percentage of variation of the oxygen solubility in the wastewater was estimated as 2.68% (Eq. A2.9) meaning that this is the percentage of extra oxygen that has to be added. This value was calculated for the concentration of 4,480 mg TDS/L, considering that compounds other than salts did not contribute to this concentration significantly. The percentage of variation of the K_{La} for a concentration of 1,920 (Case A) and 4,48 (Case B) mg TDS/L represents an increase in oxygen transfer of 5.2% according to Eq. A2.10. Thus, considering both percentages, the oxygen supplied increases by 2.52% which is the percentage of reduction of oxygen supply required.

7.4.2 Comparison of case studies

According to the calculations described in Annex 2 the following results were obtained for the case studies evaluated. The data used for the calculations will not allow for a direct analysis of the obtained costs, and percentages will be determined to compare the two cases and alternatives (Table 7.3).

The energy consumed for aeration of the secondary treatment decreased by 1.44% in Case B compared to Case A, as the oxygen solubility decreased but it was compensated by the oxygen transfer increase due to the increase of salinity (Figure A2.1). The most relevant energy-consuming activity is the functioning of the MBR and the RO systems. The energy consumption for the MBR contributes to approximately 44.5% of the total energy of Case B when the effluent is produced for discharge. In addition in Case B when reclaimed water is produced for reuse the MBR contributes to approximately 22.6% and the reverse osmosis to approximately 49% of the total energy consumption.

The SRT did not need to be modified for Case B as the concentration of salts was not sufficiently high to negatively affect the biomass activity. For this reason, the mass per day of solids produced in the activated sludge system was the same in both cases. Thus, the solid management costs remain the same.

With respect to the discharge fee (Annex 2, Eq. A2.8) the obtained amount for the WWTP example accounts for approximately 9–12% of the costs if the composition of the discharge effluent complies with the limits of the Directive. In the case where the reclaimed water is sold for agriculture irrigation the revenue obtained covers the expenses of the energy consumed and solid management in both cases. In Case A significant benefits could be achieved. If the personnel and maintenance costs are added, as 40% of the total costs, the revenue from the reclaimed water sales would cover all the costs in Cases A and B.

Table 7.3 Summary of the main results obtained from the calculations.

		Case A		Case B			
		A1	**A2**	**B1**		**B2**	
		Discharge	**Reuse**	**Discharge**		**Reuse**	
SRT	**d**	**18**	**25**	**18**	**25**	**18**	**25**
Energy consumption							
Aeration energy	kWh/y	2,180,280	2,206,819	2,149,252	2,175,122	2,149,252	2,175,122
Pumping energy	kWh/y	708,100		751,535		751,535	
MBR energy	kWh/y	–		2,324,336		2,324,336	
Reverse osmosis	kWh/y	–		–		5,036,061	
TOTAL ENERGY	kWh/y	2,888,380	2,914,919	5,225,123	5,250,993	10,261,183	10,287,054
Operational costs							
Energy	€/y	317,722	320,641	574,764	577,609	1,128,730	1,131,576
Solid management	€/y	117,139	108,371	117,139	108,371	117,139	108,371
TOTAL COSTS	€/y	724,768	715,020	1,153,171	1,143,301	2,076,449	2,066,578
Discharge fee							
WWTP example	€/y	58,503		62,092			
Water reuse benefits							
Reclaimed water	€/y	2,190,000				2,324,336	

The benefits from selling the reclaimed water increase by 6% in Case B compared to Case A. This revenue is of relevance considering the large costs of the RO treatment to produce reclaimed water suitable for reuse. If selling the reclaimed water is not feasible, the operational costs of the treatment are too high and might impede the implementation of this treatment.

The obtained results will change if the costs of the different items and the selling price of the water are adjusted to other values.

7.5 CONCLUSIONS

After the evaluation performed the following conclusions can be drawn:

The production of reclaimed water for agriculture irrigation is feasible from WWTPs with saline wastewater, but it involves the implementation of technologies specifically devoted to adjusting the required quality. Membrane technologies are by far the most commonly employed technologies to guarantee the saline conditions of the reclaimed effluents.

The energy required to operate the membrane systems represents an increment in the energy consumption of the plant compared to the base study Case A of approximately 22% in the case of the ultrafiltration system and of 49% associated with the reverse osmosis implemented for salt removal.

In the case where the regenerated water is sold for 0.3 €/m^3, in Cases A and B the amount received (2,190,000 and 2,324,336 €/y, respectively) would be enough to compensate for the operational expenses incurred including energy consumption, solid management, personnel and maintenance. Thus, it is of great importance not only to establish appropriate norms to regulate the price of reclaimed water for reuse in agriculture but also to define the concentration limits of the different pollutants.

REFERENCES

Abou-Elela S. I., Kamel M. M. and Fawzy M. E. (2010). Biological treatment of saline wastewater using a salt-tolerant microorganism. *Desalination*, **250**, 1–5.

Cadagua and Ferrovial-Agromán S.A. (2007). Plant Report: Remodelling of San Pedro del Pinatar Wastewater Treatment Plant in Murcia. Infoenviro, June, 19–33.

Candela L., Fabregat S., Josa A., Suriol A., Mas J. and Vigues N. (2007). Assessment of soil and groundwater impacts by treated urban wastewater reuse. A case study: application in a golf course (Girona, Spain). *Science of the Total Environment*, **374**, 26–35.

Carrera L., Springer F., Lipeme-Kouyi G. and Buffiere P. (2016). A review of sulphide emissions in sewer networks: overall approach and systemic modelling. *Water Science and Technology*, **73**(6), 1231–1242.

Chen G. H., Chui H. K., Wong C. L., Tang D. T. W., Lu H., Jiang F. and van Loosdrecth M. C. M. (2012). An innovative triple water supply system and an novel SANI® process to alleviate water shortage and pollution problem for water-scarce coastal areas in China. *Journal of Water Sustainability*, **2**(2), 121–129.

Chen Y., He H., Liu H., Li H., Zeng G., Xia X. and Yang C. (2018). Effect of salinity on removal performance and activated sludge characteristics in sequencing batch reactors. *Bioresource Technology*, **249**. 890–899.

Churchill P. and Elmer D. (1999). Hydrogen sulphide odor control in wastewater collection systems. *NEWEA Journal*, **33**(1), 57–63.

Duan J., Wang J., Graham N. and Wilson F. (2002). Coagulation of humic acid by aluminium sulphate in saline water conditions. *Desalination*, **150**, 1–14.

Eurostat (2019). See https://ec.europa.eu/eurostat/statistics-explained/index.php?title=File:Electricity_prices_for_non-household_consumers,_first_half_2020_(EUR_per_kWh)_v4.png (accessed 24 February 2021).

FAO (1985). Water quality for agriculture. FAO Irrigation and Drainage paper 29, Rev. 1, FAO, Rome.

Flood J. F. and Cahoon L. B. (2011). Risks to coastal wastewater collection systems from sea-level rise and climate change. *Journal of Coastal Research*, **27**, 652–660.

Ganigué R., Jiang G., Liu Y., Sharma K., Wang Y.-C., Gonzalez J., Nguyen T. and Yuan Z. (2018). Improved sulfide mitigation in sewers through on-line control of ferrous salt dosing. *Water Research*, **135**, 302–310.

Ge H., Zhang L., Batstone D. J., Keller J. and Yuan Z. (2013). Impact of iron salt dosage to sewers on downstream anaerobic sludge digesters: sulfide control and methane production. *Journal of Environmental Engineering*, **139**(4), 594–601.

Ghalavand Y., Hatamipour M. S. and Rahimi A. (2015). A review on energy consumption of desalination processes. *Desalination and Water Treatment*, **54**, 1526–1541.

Guerreiro R. C. S., Jerónimo E., Luz S., Pinheiro H. M. and Prazeres A. R. (2020). Cheese manufacturing wastewater treatment by combined physicochemical processes for reuse and fertilizaer production. *Journal of Environmental Management*, **264**, 110470.

He H., Chen Y., Li X., Cheng Y., Yang C. and Zeng G. (2017). Influence of salinity on microorganisms in activated sludge processes: A review. *International Biodeterioration and Biodegradation*, **119**, 520–527.

Henze M., Gujer W., Mino T. and van Loosdrecht M. (2002). Activated sludge models ASM1, ASM2, ASM2d and ASM3. IWA Publishing in its Scientific and Technical Report series, London, UK.

Hilal N., Al-Abri M., Moran A. and Al-Hinai H. (2008). Effects of heavy metals and polyelectrolytes in humic substance coagulation under saline conditions. *Desalination*, **220**, 85–95.

Hochstrat R., Wintgens T., Mellin T. and Jeffrey P. (2005). Wastewater reclamation and reuse in Europe – a model-based potential estimation. *Water Supply*, **5–1**, 67–75.

Jiang G., Sun X., Keller J. and Bond P. L. (2015). Identification of controlling factors for the initiation of corrosion of fresh concrete sewers. *Water Research*, **80**, 30–40.

Jiang G., Zhou M., Chiu T. H., Sun X., Keller J. and Bond P. L. (2016). Wastewater enhanced microbial corrosion of concrete sewers. *Environmental Science and Technology*, **50**(15), 8084–8092.

Kulandaivelu J., Shrestha S., Khan W., Dwyer J., Steward A., Bell L., Mcphee P., Smith P., Hu S., Yuan Z. and Jiang G. (2020). Full-scale investigation of ferrous dosing in sewers and a wastewater treatment plant for multiple benefits. *Chemosphere*, **250**, 126221.

Lefebvre O. and Moletta R. (2006). Treatment of organic pollution in industrial saline wastewater: A literature review. *Water Research*, **40**(20), 3671–3682.

Lens P. N. L., Visser A., Janssen A. J. H., Hulshoff Pol L. W. and Lettinga G. (1998). Biotechnological treatment of sulfate-rich wastewaters. *Critical Review in Environmental Science and Technology*, **28**(1), 41–88.

Leung R. W. K., Li D. C. H., Yu W. K., Chui H. K., Lee T. O., van Loosdrecht M. C. M. and Ghen G. H. (2012). Integration of seawater and grey water reuse to maximize alternative water resource for coastal areas: the case of the Hong Kong International Airport. *Water Science and Technology*, **65**(3), 410–417.

Levine A. D. and Asano T. (2004). Recovering sustainable water from wastewater. *Environmental Science and Technology*, **38**(11), 201A–208A.

Li X. Z., Luk S. F. and Tang S. L. (2005). Sustainability of toilet flushing water supply in Hong Kong. *Water and Environmental Journal*, **19**(2), 85–90.

Li X., Kappler U., Jiang G. and Curtis T. P. (2017). The ecology of acidophilic microorganisms in the corroding concrete sewer environment. *Frontiers in Microbiology*, **8**(683), 1–16.

Linarić M., Markić M. and Sipos L. (2013). High salinity wastewater treatment. *Water Science and Technology*, **68**(6), 1400–1405.

Liu X., Dai J., Wu D., Jiang F., Chen G., Chui H.-K. and van Loosdrecht M. C. M. (2016). Sustainable application of a novel water cycle using seawater for toilet flushing. *Engineering*, **2**, 460–469.

Lopez A., Pollice A., Laera G., Lonigro A. and Rubino P. (2010). Membrane filtration of municipal wastewater effluents for implementing agricultural reuse insouthern Italy. *Water Science and Technology*, **625**, 1121–1128.

Mateju V., Cizinska S., Krejci J. and Janoch T. (1992). Biological water denitrification - A review. *Enzyme Microbial Technology*, **14**, 170–183.

Metcalf and Eddy (2003). Wastewater Engineering: Treatment and Reuse, 4th edn, McGraw-Hill, New York.

Metcalf and Eddy (2014). Wastewater Engineering: Treatment and Resource Recovery. 5th Edition, McGraw-Hill, New York.

Moussa M. S., Sumanasekera D. U., Ibrahim S. H., Lubberding H. J., Hooijmans C. M., Gijzen H. J. and van Loosdrecht M. C. M. (2006). Long term effects of salt on activity, population structure and floc characteristics in enriched bacterial cultures of nitrifiers. *Water Research*, **40**, 1377–1388.

Nielsen P. H., Raunkjaer K., Norsker N. H., Jensen A. and Hvitved-Jacobsen T. (1992). Transformations of wastewater in sewer systems – a review. *Water Science and Technology*, **25**(6), 17–31.

Pedrero F., Alarcón J. J., Nicolás E. and Mounzer O. (2012). Influence of irrigation with saline reclaimed waer on young grapfruits. *Desalination and Water Treatment*, **51**, 2488–2496.

Pedrero F., Camposeo S., Pace B., Cefola M. and Vivaldi G. A. (2018). Use of reclaimed wastewater on fruit quality of nectarine in Southern Italy. *Agricultural Water Management*, **203**, 186–192.

Quinn J. J., Kracht W., Gomez C. O., Gagnon C. and Finch J. A. (2007). Comparing the effect of salts and frother (MIBC) on gas dispersion and froth properties. *Minerals Engineering*, **20**, 1296–1302.

Rhoades J. D., Kandiah A. and Mashali A. M. (1992). The use of saline waters for crop production. Food and Agricultures Organization of the United Nations, Rome, Italy.

Salehin S., Kulandaivelu J., Rebosura M., Khan W., Wong R., Jiang G., Smith P., McPhee P., Howard C., Sharma K., Keller J., Donose B. C., Yuan Z. and Pikaar I. (2019). Opportunities for reducing coagulants usage in urban water management: the Oxley Creek Sewage Collection and Treatment System as an example. *Water Research*, **165**, 114996.

Salvadó H., Mas M., Menéndez S. and Gracia M. P. (2001). Effects of shock loads of salt on protozoan communities of activated sludge. *Acta Protozoologica*, **40**, 177–185.

Tan X., Acquah I., Liu H., Li W. and Tan S. (2019). A critical review on saline wastewater treatment by membrane bioreactor (MBR) from a microbial perspective. *Chemosphere*, **220**, 1150–1162.

Tang S. L., Yue Derek P. T. and Li X. Z. (2006). Comparison of engineering costs of raw freshwater, reclaimed water and seawater for toilet flushing in Hong Kong. *Water and Environmental Journal*, **20**, 240–247.

Wang Y., Gao B. Y., Xu X. M. and Xu W. Y. (2010). The effect of total hardness and ionic strength on the coagulation performance and kinetics of aluminum salts to remove humic acid. *Chemical Engineering Journal*, **160**, 150–156.

Wang H. T., Ye Y. Y., Qi J., Li F. T. and Tang Y. L. (2013). Removal of titanium dioxide nanoparticles by coagulation: effects of coagulants, typical ions, alkalinity and natural organic matters. *Water Science and Technology*, **68**, 1137–1143.

Werzernak C. T. and Gannon J. J. (1967). Oxygen-nitrogen relationships in autotrophic nitrification, *Applied Microbiology*, **15**(5), 1211–1215.

Xie K., Xia S., Song J., Li J., Qiu L., Wang J. and Zhang S. (2014). The effect of salinity on membrane fouling characteristics in an intermittengly aerated membrane bioreactor. *Journal of Chemistry*, **2014**(6–7), 1–7.

Zannotti M. and Giovannetti R. (2015). Kinetic evidence for the effect of salts on the oxygen solubility using laboratory prototype aeration system. *Journal of Molecular Liquids*, **211**, 656–666.

Zarzo D., Campos E. and Terrero P. (2013). Spanish experience in desalination for agriculture. *Desalination and Water Treatment*, **51**(1–3), 53–66.

Annex 1

Calculations case study: Treatment of fish-canning effluents

A1.1 SCENARIO A

Table A1.1 Results of the composition of the different flows in the treatment train of Scenario A.

Parameter	Unit	Mixed Wastewater	Sieving	DAF	Coag-floc
Total suspended solids	mg/L	332.3	249.2	179.4	17.9
Volatile suspended solids	mg/L	259.7	194.7	140.2	14.0
Total COD	mg/L	1,664.0	1,503.4	1,082.5	108.2
Soluble COD	mg/L	1,021.6	868.4	607.9	182.4
Greases	mg/L	375.8	263.1	52.6	26.3
Volatile fatty acids	mg COD/L	207.2	0.0	0.0	0.0
Ammonia	mg N/L	31.2	31.2	31.2	31.2
Total nitrogen	mg N/L	68.8	68.8	68.8	68.8
Phosphate	mg P/L	8.5	8.5	8.5	8.5
Chloride	mg Cl^-/L	8,200.0	8,200.0	8,200.0	8,200.0
Sulphate	mg SO_4^{-2}/L	1,086.0	1,086.0	1,086.0	1,086.0
Sodium	mg Na^+/L	4,404.0	4,404.0	4,404.0	4,404.0

Both streams are treated together ($Q = 88$ m^3/d).

Author(s): Anuska Mosquera Corral, Ángeles Val del Río and José Luis Campos Gómez
doi: 10.2166/9781789060645_0173

A1.2 SCENARIO B

Table A1.2 Results of the composition of the different flows in the treatment train of Scenario B.

Parameter	Unit	High load	Sieving	DAF	AS
Total suspended solids	mg/L	2,370	1,777.5	1,279.8	128.0
Volatile suspended solids	mg/L	1,070	802.5	577.8	57.8
Total COD	mg/L	10,800	9,925.0	7,146.0	714.6
Soluble COD	mg/L	7,300	6,205.0	4,343.5	434.4
Greases	mg/L	1,600	1,120.0	224.0	224.0
Ammonia	mg N/L	560	560.0	560.0	560.0
Total nitrogen	mg N/L	1,020	1,020.0	1,020.0	1,020.0
Phosphate	mg P/L	130	130.0	130.0	130.0
Chloride	mg Cl^{-}/L	4,210	4,210.0	4,210.0	4,210.0
Sulphate	mg SO_4^{-2}/L	250	250.0	250.0	250.0
Sodium	mg Na^{+}/L	2,770	2,770.0	2,770.0	2,770.0
Parameter	**Unit**	**Low load**	**Sieving**	**DAF**	**Coag-floc**
Total suspended solids	mg/L	225.0	168.8	121.5	12.2
Volatile suspended solids	mg/L	217.0	162.8	117.2	11.7
Total COD	mg/L	1,183.2	1,060.2	763.3	76.3
Soluble COD	mg/L	691.2	587.5	411.3	123.4
Greases	mg/L	311.4	218.0	43.6	21.8
Ammonia	mg N/L	3.4	3.4	3.4	3.4
Total nitrogen	mg N/L	18.7	18.7	18.7	18.7
Phosphate	mg P/L	2.1	2.1	2.1	2.1
Chloride	mg Cl^{-}/L	8,410.0	8,410.0	8,410.0	8,410.0
Sulphate	mg SO_4^{-2}/L	1,130.0	1,130.0	1,130.0	1,130.0
Sodium	mg Na^{+}/L	4,490.0	4,490.0	4,490.0	4,490.0

Each stream is treated separately, and the global effluent is a mixture of the streams produced in each treatment line (high load wastewater $Q = 4.4$ m^3/d and low load wastewater $Q = 83.6$ m^3/d).

A1.3 SCENARIO C

Table A1.3 Results of the composition of the different flows in the treatment train of Scenario C.

Parameter	Unit	High Load	Sieving	DAF	AD
Total suspended solids	mg/L	2,370	1,777.5	1,279.8	128.0
Volatile suspended solids	mg/L	1,070	802.5	577.8	57.8
Total COD	mg/L	10,800	9,925.0	7,146.0	714.6
Soluble COD	mg/L	7,300	6,205.0	4,343.5	868.7
Greases	mg/L	1,600	1,120.0	224.0	22.4
Volatile fatty acids	mg COD/L	3,300	0.0	0.0	0.0
Ammonia	mg N/L	560	560.0	560.0	560.0
Total nitrogen	mg N/L	1,020	1,020.0	1,020.0	1,020.0
Phosphate	mg P/L	130	130.0	130.0	130.0
Chloride	mg Cl^-/L	4,210	4,210.0	4,210.0	4,210.0
Sulphate	mg SO_4^{-2}/L	250	250.0	250.0	250.0
Sulphide	mg H_2S/L	–	–	–	0.054
Sodium	mg Na^+/L	2,770	2,770.0	2,770.0	2,770.0
Parameter	**Unit**	**Low load**	**Sieving**	**DAF**	**Coag-floc**
Total suspended solids	mg/L	225.0	168.8	121.5	12.2
Volatile suspended solids	mg/L	217.0	162.8	117.2	11.7
Total COD	mg/L	1,183.2	1,060.2	763.3	76.3
Soluble COD	mg/L	691.2	587.5	411.3	123.4
Greases	mg/L	311.4	218.0	43.6	21.8
Volatile fatty acids	mg COD/L	44.5	0.0	0.0	0.0
Ammonia	mg N/L	3.4	3.4	3.4	3.4
Total nitrogen	mg N/L	18.7	18.7	18.7	18.7
Phosphate	mg P/L	2.1	2.1	2.1	2.1
Chloride	mg Cl^-/L	8,410.0	8,410.0	8,410.0	8,410.0
Sulphate	mg SO_4^{-2}/L	1,130.0	1,130.0	1,130.0	1,130.0
Sulphide	mg H_2S/L	–	–	–	–
Sodium	mg Na^+/L	4,490.0	4,490.0	4,490.0	4,490.0

Each stream is treated separately, and the global effluent is a mixture of the streams produced in each treatment line (high load wastewater Q = 4.4 m^3/d and low load wastewater Q = 83.6 m^3/d).

A1.4 SCENARIO D

Table A1.4 Results of the composition of the different flows in the treatment train of Scenario D.

Parameter	Unit	High Load	Sieving	DAF	AD	AS+N/DN
Total suspended solids	mg/L	2,370	1,777.5	1,279.8	128.0	12.8
Volatile suspended solids	mg/L	1,070	802.5	577.8	57.8	5.8
Total COD	mg/L	10,800	9,925.0	7,146.0	714.6	71.5
Soluble COD	mg/L	7,300	6,205.0	4,343.5	868.7	86.9
Greases	mg/L	1,600	1,120.0	224.0	22.4	22.4
Ammonia	mg N/L	560	560.0	560.0	560.0	56.0
Total nitrogen	mg N/L	1,020	1,020.0	1,020.0	1,020.0	102.0
Phosphate	mg P/L	130	130.0	130.0	130.0	130.0
Chloride	mg Cl^-/L	4,210	4,210.0	4,210.0	4,210.0	4,210.0
Sulphate	mg SO_4^{-2}/L	250	250.0	250.0	250.0	250.0
Sulphide	mg H_2S/L	–	–	–	0.054	0.0
Sodium	mg Na^+/L	2,770	2,770.0	2,770.0	2,770.0	2,770.0
Parameter	**Unit**	**Low Load**	**Sieving**	**DAF**	**Coag-floc**	
Total suspended solids	mg/L	225.0	168.8	121.5	12.2	
Volatile suspended solids	mg/L	217.0	162.8	117.2	11.7	
Total COD	mg/L	1183.2	1,060.2	763.3	76.3	
Soluble COD	mg/L	691.2	587.5	411.3	123.4	
Greases	mg/L	311.4	218.0	43.6	21.8	
Ammonia	mg N/L	3.4	3.4	3.4	3.4	
Total nitrogen	mg N/L	18.7	18.7	18.7	18.7	
Phosphate	mg P/L	2.1	2.1	2.1	2.1	

(Continued)

Table A1.4 Results of the composition of the different flows in the treatment train of Scenario D (*Continued*).

Parameter	Unit	High Load	Sieving	DAF	AD	AS+N/DN
Chloride	mg Cl^{-}/L	8,410.0	8,410.0	8,410.0	8,410.0	
Sulphate	mg SO_4^{-2}/L	1,130.0	1,130.0	1,130.0	1,130.0	
Sulphide	mg H_2S/L	–	–	–	–	
Sodium	mg Na^{+}/L	4,490.0	4,490.0	4,490.0	4,490.0	

Each stream is treated separately, and the global effluent is a mixture of the streams produced in each treatment line (high load wastewater Q = 4.4 m^3/d and low load wastewater Q = 83.6 m^3/d).

A1.5 BASE DATA FOR COSTS CALCULATION

Table A1.5 Typical energy consumption of various treatment processes on wastewater treatment according to Metcalf and Eddy (2014).

	Average kWh/m^3
Wastewater influent pumping	0.0385
Screens	0.0004
Activated sludge for BOD removal	0.14
Activated sludge with nitrification/denitrification	0.23
Return sludge pumping	0.0155
Secondary settling	0.0035
Dissolved air floatation	0.035
Mesophilic anaerobic digestion (primary plus waste activated sludge)	0.1265

Values were chosen as the average of the interval provided in the document.

REFERENCE

Metcalf and Eddy (2014). Wastewater Engineering: Treatment and Resource Recovery. 5th Edition, McGraw-Hill, New York.

Annex 2

Calculations case study: Urban wastewater treatment plant with saline intrusions

A2.1 COMPOSITION OF THE SEAWATER

Table A2.1 Composition of the artificial seawater (ASTM D 1141-98, Formula A).

Salt	Concentration (g/L)	Salt	Concentration (g/L)
NaCl	24.53	KBr	0.101
$MgCl_2$	5.2	H_3BO_3	0.027
Na_2SO_4	4.09	$SrCl_2$	0.025
$CaCl_2$	1.16	NaF	0.003
KCl	0.695	Water	988.968
$NaHCO_3$	0.201		

A2.2 EQUATIONS USED FOR TECHNICAL CALCULATIONS

The calculations performed in the case studies to determine the following parameters were summarized from Metcalf & Eddy (2013). The resulting values were used to perform the evaluation of the cases studies.

Author(s): Anuska Mosquera Corral, Ángeles Val del Río and José Luis Campos Gómez
doi: 10.2166/9781789060645_0179

A2.2.1 Net waste activated sludge produced ($P_{X,VSS}$) in the activated sludge system

For this calculation, the biomass produced from the nitrification and denitrification processes, the heterotrophic biomass decay and the solids forming as inert solids in the wastewater were considered.

$$P_{X,VSS} = Q_0 \cdot \left((1 + f_d \cdot k_d \cdot \theta_C) \cdot \frac{Y_{X/COD} \cdot (S_0 + X_{S,0} \cdot 0.4 - S_e)}{1 + k_d \cdot \theta_C} + \frac{Y_{X,N} \cdot S_N}{1 + k_{dN} \cdot \theta_C} + \frac{X_{i0}}{1.55} \right) \tag{A2.1}$$

where:

$P_{X,VSS}$ = net waste activated sludge produced each day, kg VSS · d
f_d = fraction of cell mass remaining as cell debris, kg VSS/kg VSS
k_d = endogenous decay coefficient for heterotrophic organisms, kg VSS/kg VSS · d
θ_C = solids retention time, d
Q_0 = influent flow, m^3/d
$Y_{X/S}$ = observed yield coefficient, kg VSS/kg (COD or NH_4^+-N) removed
S_0 = influent readily biodegradable COD concentration, kg COD/m^3
S_e = effluent readily biodegradable COD concentration, kg COD/m^3
S_N = ammonium nitrogen concentration oxidized to nitrate, kg N/m^3
k_{dN} = endogenous decay coefficient for nitrifying organisms, kg VSS/kg VSS · d
X_{i0} = influent particulate non-biodegradable COD concentration, kg COD/m^3
$X_{S,0}$ = influent slowly biodegradable particulate COD concentration, kg COD/m^3

A2.2.2 Activated sludge wasted ($W_{X,TSS}$)

The mass of activated sludge wasted to maintain the SRT value was calculated from the $P_{X,VSS}$ converted to TSS units by dividing by a factor of 0.75 kg VSS/kg TSS and then subtracting from this amount the mass of solids leaving the system (after the membrane bioreactor).

$$W_{X,TSS} = \frac{P_{X,VSS}}{0.75} - Q_e \cdot X_{TSS,e} \tag{A2.2}$$

where:

$W_{X,TSS}$ = activated sludge wasted each day, kg TSS/d
Q_e = effluent flow, m^3/d
$X_{TSS,e}$ = effluent activated sludge concentration, kg TSS/m^3

A2.2.3 Total solids wasted in the WWTP ($TW_{X,TSS}$)

The total solids wasted in the plant were equal to the activated sludge wasted in the activated sludge system plus the fraction of solids separated in the pre-treatments, which was estimated as 60% of the solids entering the facility.

$$TW_{X,TSS} = W_{X,TSS} + Q_0 \cdot X_{S,0} \cdot 0.6 \tag{A2.3}$$

where:

$TW_{X,TSS}$ = total solids wasted each day in the WWTP, kg TSS/d

A2.2.4 Total oxygen required for aerobic biological processes

$$R_{O2} = \left(Q_0 \cdot (S_0 - S) - 1.42 \cdot (1 + f_d \cdot k_d \cdot \theta_C) \cdot Q_0 \cdot \frac{Y_{X/COD} \cdot (S_0 - S)}{1 + k_d \cdot \theta_C} + 4.33 \cdot Q_0 \cdot S_N \right) \tag{A2.4}$$

where:

R_{O2} = total oxygen required, kg O_2/d

Werzemak and Gannon (1967) found that the actual total oxygen consumption was 4.33 g O_2/g N with 3.22 g O_2/g N used for ammonia oxidation and 1.11 g O_2/g N used for nitrite oxidation.

The values chosen for the kinetic parameters (chosen at 20°C) corresponding to heterotrophic and nitrifying bacteria are shown in Table A2.2.

Table A2.2 Values of the kinetic parameters (20°C) chosen for the calculations (Metcalf & Eddy, 2013).

Coefficient	Unit	Value
$Y_{X/COD}$	kg VSS/kg COD removed	0.40
$Y_{X/N}$	kg VSS/kg NH_4^+-N removed	0.12
f_d	kg VSS/kg VSS	0.15
k_d	kg VSS/kg VSS · d	0.12
k_{dN}	kg VSS/kg VSS · d	0.08

$Y_{X/N}$: observed yield coefficient, kg VSS/kg NH_4^+-N removed.

A2.2.5 Energy consumption due to aeration and pumping activities

$$EnC = R_{O2} \cdot \alpha + Q_0 \cdot (\beta_1 + \beta_2 \cdot R) \tag{A2.5}$$

where:

EnC = energy consumption due to aeration and pumping activities (kWh/d)
α = power used for mass of oxygen transferred fixed as 1 kWh/kg O_2
Q_{pe} = wastewater flow rate corresponding to 1 person equivalent fixed as 0.154 m^3/d
β_1 = power required to pump the water per day fixed as 0.045 kWh/m^3
β_2 = power required to pump the recycled sludge per day fixed as 0.013 kWh/m^3
R = recycling ratio

A2.2.6 WWTP discharge fee

The annual fee for discharge (DF) to the sea is calculated according to the Sectorial Environmental Authorization corresponding to the example WWTP (in San Pedro del Pinatar) as follows using Eqs. (A2.6), (A2.7) and (A2.8):

$$PL = SSL + OML \tag{A2.6}$$

$$PU = \frac{DPL}{RPL} \tag{A2.7}$$

where:

PL = pollution load (kg/y)
SSL = load of suspended solids (kg/y)
OML = oxidable matter load equivalent to 2/3 of the COD (kg/y)
PU = pollution units that have a price of 6,000 €
DLP = discharged pollution load
RPL = Reference pollution load equal to 53,655 kg (example WWTP)

From the previous results the fee associatfed with effluent discharge is calculated using Eq. (A2.8):

$$DF = PU \cdot 6{,}000 \cdot k \tag{A2.8}$$

where:

DF = Discharge fee (€/y)
k = factor of correction depending on the type of wastewater. In this case it is equal to 1 for urban wastewater discharged to the sea through a discharge pipe.

A2.3 DEPENDENCY OF OXYGEN SOLUBILITY

The effect of salt concentration on the oxygen solubility and oxygen mass transfer coefficient (K_{La}) was considered for the estimation of the energy needed for oxygen supply in the secondary treatment. The oxygen solubility at 20°C decreased by 5.5% in the case of water without salt compared to a sample of water containing 9 g/L of salt.

The correlation of the percentage of variation, with respect to the case with no salt, from the data of Figure A2.1 was adjusted as a function of the salt concentration (Eq. A2.9) to a variation of 2.68% which corresponds to the increment of oxygen needed in the activated sludge system.

$$\Delta S_{O2}(\%) = 0.55 \cdot S_{salt} + 0.22 \quad \text{(A2.9)}$$

Additionally the K_{La} estimated according to Zanotti and Giovannetti (2015) using Eq. (A2.10) is:

$$K_{La} = 0.5267 \cdot \ln(S_{salt}) + 7.8779 \quad \text{(A2.10)}$$

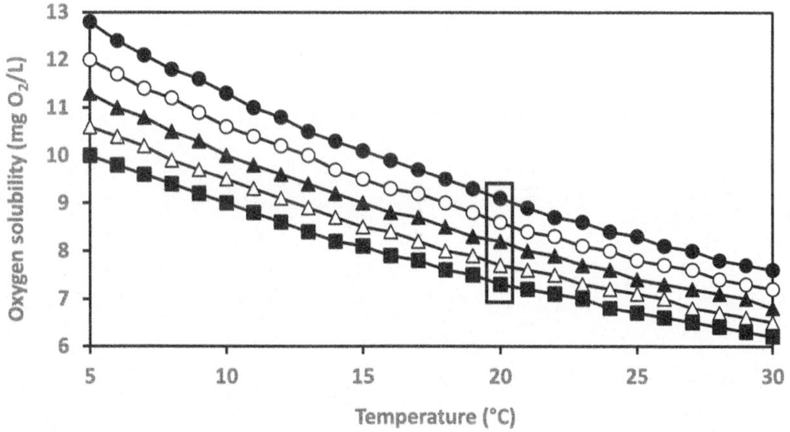

Figure A2.1 Concentration of dissolved oxygen (mg O_2/L) as a function of temperature and salt concentrations of 0 g/L (●), 9 g/L (○), 18 g/L (▲), 27 g/L (△) and 36 g/L (■). (Data from EPA-842-B-06-003.)

REFERENCES

EPA-842-B-06-003 (2006). Chapter 9 of the volunteer estuary monitoring manual, A Methods Manual, 2nd edn. https://www.epa.gov/sites/production/files/2015-09/documents/2009_03_13_estuaries_monitor_chap9.pdf (last access 19/01/2021).

Metcalf & Eddy Inc., Tchobanoglous G., Burton F. L., Tsuchihashi R. and Stensel H. D. (2013). Wastewater engineering: Treatment and resource recovery (5th ed.). McGraw-Hill Professional.

Werzernak C. T. and Gannon J. J. (1967). Oxygen-nitrogen relationships in autotrophic nitrification. *Applied Microbiology*, **15**(5), 1211–1215.

Zanotti M. and Giovannetti R. (2015). Kinetic evidence for the effect of salts on the oxygen solubility using laboratory prototype aeration system. *Journal of Molecular Liquids*, **211**, 656–666.

Nomenclature

Acronym	Unit	Definition
Af		Aerobic Feeding
AF		Anaerobic Filter
AFB		Anaerobic Fluidized Bed
AGS		Aerobic Granular Sludge
AD		Anaerobic Digester
Anf		Anaerobic Feeding
AOB		Ammonium Oxidizing Bacteria
AOP		Advanced Oxidation Processes
BES		BioElectrochemical System
BOD		Biological Oxygen Demand
CAD		Central Activity Digester
CAS		Conventional Activated Sludge
CBR		Circulating Bed Reactor
C-F		Coagulation – Flocculation
COD		Chemical Oxygen Demand
CPD		Cyclobutane Pyrimidine Dimers
CST		Capillary Suction Time
CW		Constructed Wetland
CWPO		Catalytic Wet Peroxide Oxidation
DAF		Dissolved Air Floatation

(*Continued*)

Author(s): Anuska Mosquera Corral, Ángeles Val del Río and José Luis Campos Gómez
doi: 10.2166/9781789060645_0185

(*Continued*).

Acronym	Unit	Definition
DNA		DeoxyriboNucleic Acid
DOC		Dissolved Organic Carbon
DLVO		Derjaguin, Landau, Verwey, Overbeek
EC	μS/cm	Electrical Conductivity
EDR		ElectroDialysis Reversal
EPS		ExoPolySaccharides
EUMOFA		EUropean Market Observatory for Fisheries and Aquaculture products
FAO		Food and Agriculture Organization of the United Nations
FDFO		Fertilizer Drawn Forward Osmosis
FO		Forward Osmosis
FOG		Fats, Oil and Grease
GAO		Glycogen Accumulating Organisms
GHG		Greenhouse Gases
HRT	d	Hydraulic Retention Time
HB		Hydroxybutyrate
HS		Humic Substances
HV		Hydroxy valerate
ISABR		Immobilized Sludge Anaerobic Baffled Reactor
LCA		Life Cycle Assessment
LP		Low Pressure
LR	$kg/(m^3 \cdot d)$	Loading Rate
MBBR		Moving Bed Biofilm Reactors
MBR		Membrane Bioreactor
MDC		Microbial Desalination Cell
ML		Medium Pressure
MMC		Microbial Mixed Culture
NF		Nanofiltration
NLR	$kg\ N/(m^3 \cdot d)$	Nitrogen Loading Rate
NOB		Nitrite Oxidizing Bacteria
OLR	$kg\ COD/(m^3 \cdot d)$	Organic Loading Rate
PAC		Polyalanine Chloride
PAO		Phosphorus Accumulating Organisms

(*Continued*)

(*Continued*).

Acronym	Unit	Definition
PBBR		Packed-Bed Biofilm Reactor
PHA		PolyHydroxyAlkanoates
PM		Particulate matter
PN-AMX		Partial Nitritation – Anammox process
PPB		Purple Phototrophic Bacteria
PVC		PolyVinylChloride
R		Recycling ratio
RO		Reverse Osmosis
SANI		Sulphate reduction Autotrophic denitrification and Nitrification Integrated
SAR		Sodium Absorption Ratio
SBAR		Sequencing Batch Airlift Reactor
SBR		Sequencing Batch Reactor
Si	mg COD/L	Non-biodegradable soluble COD fraction
SP		Sulphated Polysaccharides
SRB		Sulphate Reducing Bacteria
SRF		Specific Resistance in Filtration
SRT	d	Sludge Retention Time
Ss	mg COD/L	Readily biodegradable soluble COD fraction
TDS	mg/L	Total Dissolved Solids
TKN		Total Kjeldahl Nitrogen
TN		Total Nitrogen
TPH		Total Petroleum Hydrocarbons
TRO		Tubular Reverse Osmosis
TSS	g/L	Total Suspended Solids
TSSE		Temperature Swing Solvent Extraction
TWS		Triple Water Supply
UAF		Upflow Anaerobic Filter
UF		UltraFiltration
UFM		UltraFiltration Module
UASB		Upflow Anaerobic Sludge Blanket
UV		UltraViolet
VFA		Volatile Fatty Acids

(*Continued*)

(*Continued*).

Acronym	Unit	Definition
VS		Volatile Solids
VSS	g/L	Volatile Suspended Solids
WAS		Waste Activated Sludge
WWTP		WasteWater Treatment Plant
Xi	mg COD/L	Non-biodegradable particulate COD fraction
Xs	mg COD/L	Slowly biodegradable particulate COD fraction
ZLD		Zero Liquid Discharge

IWA Publishing's authorised EU representative for General Product Safety Regulations is Diane D'Arras, 15 rue Duret, 75116 Paris, France, e-mail: safety@iwap.co.uk.

Printed and bound by CPI Group (UK) Ltd, Croydon, CR0 4YY
16/06/2026
02140394-0001